T0262420

IEE Computing Series 1

Series Editors: Dr. D.A.H. Jacobs, M. Sage,
Prof S.L. Hurst, Dr. B. Carré
and Prof I. Sommerville

Semi-Custom IC Design and VLSI

Other volumes in this series

Semi-Custom IC Design and VLSI

Edited by P. J. Hicks

Peter Peregrinus Ltd
on behalf of the Institution of Electrical Engineers

Published by: Peter Peregrinus Ltd., London, United Kingdom

© **1983: Peter Peregrinus Ltd.**

Reprinted 1987

All rights reserved. No part of this publication may be reproduced, stored in a retrieval system or transmitted in any form or by any means—electronic, mechanical, photocopying, recording or otherwise—without the prior written permission of the publisher.

While the author and the publishers believe that the information and guidance given in this work is correct, all parties must rely upon their own skill and judgment when making use of it. Neither the author nor the publishers assume any liability to anyone for any loss or damage caused by any error or omission in the work, whether such error or omission is the result of negligence or any other cause. Any and all such liability is disclaimed.

ISBN 0 86341 011 1

Printed in England by Short Run Press Ltd., Exeter

Contents

List of contributors

Chapters 1 and 4
P.J. Hicks
UMIST, Manchester, UK

Chapters 2 and 3
A.D. Milne
Wolfson Microelectronics
Institute, University of
Edinburgh

Chapter 5
J.R. Grierson
British Telecom Research
Laboratories, Ipswich, UK

Chapter 6
J.W. Tomkins
Plessey Research (Caswell)
Ltd.

Chapters 7 and 9
E.L. Dagless
University of Bristol, UK

Chapters 8 and 12
D.J. Kinniment
University of Newcastle upon
Tyne, UK

Chapter 10
H.G. Adshead
ICL, Manchester, UK

Chapter 11
K. Dimond
University of Kent at
Canterbury, UK

Chapter 13
M.A. Jack
University of Edinburgh,
Scotland

Chapter 14
P.B. Denyer
University of Edinburgh,
Scotland

Chapter 15
A. Hopper and J. Herbert
University of Cambridge
Computer Laboratory, UK

Preface

For twenty years or more it has been common practice for digital systems to be designed and constructed using standard "off-the-shelf" IC components. These range in complexity from small and medium scale integration (SSI and MSI) devices such as the TTL, ECL and CMOS logic families on the one hand to large scale (LSI) chips like the microprocessor on the other. With the transition that is now taking place from the LSI to the VLSI (Very Large Scale Integration) era we will soon be witnessing the emergence of 'systems on silicon'; single chips containing upwards of a million transistors. Designing chips of this complexity presents new challenges and will lead to radical changes in the activities traditionally undertaken by digital system designers. Semi-custom ICs such as gate arrays and standard cells are already enabling designers to integrate digital subsystems customised to their own specification directly onto silicon. Moreover this can be achieved without highly specialised knowledge or training in the techniques of microelectronic device design and fabrication.

In order to fully exploit the potential that micro-electronics technology has to offer it is imperative that system design engineers should become familiar with the various IC design styles that currently exist as well as be aware of possibilities for the future. The 1st IEE Vacation School on "Semi-custom IC design and VLSI" was held in July 1983 in the Department of Electrical Engineering at the University of Edinburgh with the aim of achieving some of the above objectives. Lectures on topics ranging from silicon processing technology to design automation and from gate arrays to silicon compilers were presented by specialists in each of the subject areas. The lectures are reproduced here as the chapters of this book and as such comprise a self-contained introduction to custom and semi-custom IC technology.

P.J. Hicks

Chapter 1

Introduction

P.J. Hicks

If the history of the semiconductor industry is traced back about 20 years to the early 1960's we find the earliest examples of integrated circuits starting to appear. The semiconductor manufacturers quickly realised that the economics of IC production are such that the devices must be produced in large numbers if the unit cost per device is not to be prohibitive. This is simply because the large overheads associated with each design in terms of manpower, tooling costs etc. have to be amortised over large volume sales. It was for this reason that standard ranges of IC components became established in the shape of the logic families such as Resistor-Transistor Logic (RTL), Diode-Transistor Logic (DTL) and later Transistor-Transistor Logic (TTL). Each chip within one of these families would typically contain a small number of logic gates or a flip-flop or two, and as a result could be applied widely in the design of digital systems.

During the 1960's, as the technology used to produce ICs progressively improved, it became feasible to manufacture more and more components on a single chip. Instead of just a few gates (typically a few 10's of transistors) it was now possible to integrate a logical sub-system such as a binary counter, multiplexor or even an Arithmetic and Logic Unit onto individual chips. The technology had progressed from Small Scale Integration (SSI) to Medium Scale Integration (MSI) with up to several hundred transistors now being involved. As this change took place it was noticed that the function of the MSI chips was inherently more specialised than the basic SSI gates and flip-flops. This did not really matter at the time because the functions chosen for the MSI chips were all common ones that tended to crop up again and again in digital systems - the counter is a good example.

With still further improvements in IC manufacturing technology it was possible by 1970 to produce integrated circuits containing thousands of transistors and the era of Large Scale Integration (LSI) had dawned. The degree of specialisation of function that chips of this complexity exhibited was now beginning to be something of a problem. With a few notable exceptions (digital watch and clock chips and single-chip digital voltmeters are examples) logic systems having this number of gates were so specialised that they could never be manufactured and sold in the sort of quanti-

ties that were necessary to make it an economic proposition.
To give some idea of the scale of the problem, the development costs for an LSI chip could amount to somewhere in the
region of £100K. This implies that production volumes of
the order of 100,000 units would be needed for the pricing
of the chip to be realistic.

At about this time a way out of the dilemma was forth-
coming in the shape of the microprocessor. The breakthrough
here is that a general-purpose LSI component can be produced
in high volumes because the function to be performed by the
chip isn't fixed at the time of its manufacture. Instead it
is determined at a later stage by the user who simply writes
a program for execution by the microprocessor system and
this specifies its function. Microprocessors have therefore
enabled LSI solutions to be applied to problems that did not
warrant a full-custom design.

Unfortunately, however, microprocessors are incapable
of providing solutions to all of the problems that are likely to be encountered in digital system design. In particular their sequential mode of operation means that they are
slow in performing complex tasks, and therefore inappropriate
in situations demanding high-speed processing or a large
element of parallelism. This means that in many cases there
will be no satisfactory alternative to custom-designing a
chip for a specific application - always assuming, of course,
that such a solution is economically viable.

Some of the more important advantages to be gained from
implementing a design in the form of an LSI chip are listed
below :
  (i)   reduced manufacturing costs
         -   high packing density
         -   lower system component count
         -   simplifies assembly.
  (ii)   improved system performance
         -   lower power dissipation per gate
         -   higher switching speeds.
  (iii) improved system reliability
  (iv)   greater design security.
From what has already been said it should be apparent that
there is likely to be a large demand for an IC that can
achieve LSI levels of integration without incurring the high
development costs of a full custom design. Such a device
would ideally become economic in quantities of thousands
rather than hundreds of thousands, and therefore the design
and development effort, combined with tooling costs, must be
reduced by a proportionate amount. As levels of integration
increase further towards VLSI the design effort needed for
full custom design is likely to grow drastically unless some
means can be found of radically reducing it.

The designer of a full-custom integrated circuit has
traditionally 'hand-crafted' the individual transistors and
other devices that make up his circuit. He does this by
using semiconductor device physics to predict their behaviour
before drawing out the geometric details of the devices and
their interconnections on a large scale. These will eventually be transformed into the microscopic features in the

surface of the silicon chip itself. A review of the complex sequence of wafer processing steps needed to perform this transformation is provided in chapter 2, while chapter 3 looks at the different IC technologies that are available.

It should be possible to shorten the time taken to design an integrated circuit by isolating the designer from much of the complexity of low-level chip design. By constraining the designer to use predefined components or groups of components he no longer has to be concerned with the characteristics of individual devices and instead designs at the gate or functional block level. In general the greater the number of constraints imposed the easier the task of the designer becomes. Of course the penalty that must be paid for this simplification is a loss of flexibility and a lowering of the packing density of the chip.

The concept of a semi-custom integrated circuit is based on various techniques that have been developed for achieving the objectives stated above. The most common of these techniques have been classified under the generic headings of gate arrays on the one hand and standard cells on the other. A gate array consists of a regular matrix of cells containing logic gates or components that are predefined up to the final stage of processing. This normally involves patterning one or more layers of metal to form the wires or tracks needed to connect the pre-defined components together. In this way the gate array can be configured to realise any required logic assembly. Standard cells comprise a library of predesigned functional cells that are equivalent to standard SSI and MSI logic families. Cells can be selected from the library (according to the eventual function that the chip is to perform) and arranged in parallel rows with tracking lanes in between. The wires needed to connect the cells together are then routed in metal through the tracking lanes.

There are two other techniques that are commonly grouped under the heading of semi-custom ICs and they should be mentioned here also. The first category includes fusible logic arrays such as Programmable Logic Arrays (PLA), Programmable Read Only Memories (PROM) and Programmable Array Logic (PAL). Sometimes these devices are excluded from the definition of semi-custom ICs because they are not mask-programmable. Finally, analogue component arrays are available which are capable of implementing customised analogue functions using the same basic principles as the gate array.

The semi-custom design techniques outlined above are described in detail in chapter 4. This is followed in chapter 5 by a survey of the criteria which a potential user must weigh up before deciding on the semi-custom style he should employ and his choice of supplier.

Most engineers engaged in digital system design still think primarily at the logic gate level, this being the natural 'language' for system description and hence also for the synthesis of new systems. Gate arrays and cell-based systems are both well-suited to this mode of design since the details of the underlying circuitry are to a large extent hidden behind the pre-defined logic functions. Nevertheless, an appreciation of the principles governing the operation of the circuits at the device level can help the system designer to recognise their limitations as well as appreciating their full design potential. For this reason a review of bipolar and MOS circuit design techniques has been provided in chapter 6.

Logic design is presented in chapter 7 as an exercise that can be carried out at a level one step removed from elements such as gates with their technology-dependent properties. This method, known as the Algorithmic State Machine or ASM technique, has much to recommend it since it allows the engineer to carry out the design of digital systems in an implementation-independent fashion. Various ways of implementing the ASM chart are discussed in chapter 9, including matrix logic structures such as ROM and PLA. The PLA itself is first introduced in chapter 8 as a useful "building block" approach for VLSI circuits, and an automated technique for the generation of PLAs is described.

One of the key factors that has certainly influenced the widespread acceptance of semi-custom ICs has been the availability of increasingly sophisticated yet affordable Computer-Aided Design (CAD) tools. The significance of this is twofold. Firstly, the tools for design verification, layout and checking are essential if the primary semi-custom goals of fast turnaround and first-time success are to be achieved. Secondly, the semi-custom IC manufacturers are finding it increasingly desirable to transfer a large proportion of the responsibility for the design of semi-custom circuits into the hands of their customers. This is particularly true in the case of gate arrays where the design process in many cases is still not fully automated. To support this trend several manufacturers are making available low-cost engineering workstations which can be installed on the customers premises. Running on these workstations will typically be software capable of allowing the customer/designer to access a range of CAD tools tailored to suit the particular manufacturer's range of semi-custom devices.

An overview of the present state-of-the-art in CAD and Design Automation (DA) is presented in chapter 10. Later chapters deal with various aspects of CAD and DA in greater detail : thus simulation techniques are reviewed in chapter 11 while chapter 12 examines the algorithms used for cell placement and the routing of interconnections. The important yet often neglected subjects of testing and design for testability are surveyed in chapter 13, the emphasis here being particularly directed towards the formidable problems that the testing of highly-complex VLSI chips can pose.

Looking a little way into the future of custom IC design, chapter 14 highlights some of the research that is currently being carried out into the viability of so-called silicon compilers. At the present time these are based on the concept of taking as input a high-level or procedural description of a circuit or system and delivering as output a physical layout of a chip implemented according to some pre-arranged overall floor plan. The basic structural elements or building blocks needed to construct the chip are kept in a library, although the ways in which these elements can be combined and arranged on the finished chip are no-where near as constrained as they are for conventional standard cells.

The final chapter of the book examines some practical aspects associated with implementing a logic design on a semi-custom IC and is based on experience gained from real-life applications. The role that CAD tools can play in helping the designer to avoid some of the possible pitfalls is also discussed.

In the final analysis it may no longer be very meaning-ful to discriminate between 'custom' and 'semi-custom' design, although at present there do appear to be differing views concerning the definition of these terms. Indeed, custom design performed in the traditional 'hand-crafted' manner will probably cease to be economically viable in the not-too-distant future and designers will learn to live with the constraints that need to be imposed in order to introduce automation of the design process. Whether we then choose to label the product of imposing such constraints as 'custom' or 'semi-custom' is largely immaterial, since the objective in either case must be to make the capability to design 'systems on silicon' as freely and widely available as poss-ible.

Chapter 2

Introduction to silicon fabrication

A.D. Milne

2.1 INTRODUCTION

The technology for manufacturing silicon integrated circuits has developed enormously over the past 20 years, yet many of the basic physical processes which are used today were first perfected in the early 60's. The recent developments have been in the nature of refinements of these basic processes to allow finer geometric structures to be produced with closer tolerances on the associated electrical parameters. As dimensions approach one micron ($\mu = 10^{-6}$m) fundamental limitations such as the wavelength of visible light dictate new approaches to circuit definition while increases in physical size of die and wafers require improved methods of replicating the images. The traditional method of diffusion of impurities to obtain the appropriate resistance values within the semiconductor devices, while still widely used, is being replaced at many stages of manufacture by direct ion implantation from a high energy source. This allows more accurate doping at a lower temperature, both essential features in the manufacture of complex circuits.

These developments together with those in oxidation and etching will be briefly discussed in the following sections before a full manufacturing process is described. As an indication of the future possibilities some of the new technologies which are being developed will also be described.

2.2 BASIC MATERIAL

The raw silicon material is produced by chemical manufacture as a boule of single crystal about 2m in length and from 75 - 125 mm in diameter. This is subsequently cut into wafers about ·2 - ·3 mm thick and polished to remove surface damage caused by the mechanical preparation. The quality of the wafers is extraordinarily high with impurities uniformly distributed and defined to a few parts in 10^7. The physical structure is dislocation free with only slight variations of the lattice parameters. Two growth techniques are used to provide such material; Czochralski

(CZ) which is used for the bulk of the industries requirements and Float Zoned (FZ) which provides the higher quality but is more expensive to produce and is usually limited in diameter. In the former the material is grown from a melt contained in a quartz crucible. The molten silicon corrodes the surface of the crucible which contaminates the crystal to a small degree with oxygen, whereas in float zoning the crystal is held in an inert atmosphere by surface tension while it is molten. For many applications modern CZ material is sufficient but the slight variation in resistivity across a wafer can be unacceptable for sensitive devices such as analogue and imaging devices.

The orientation of the wafer, that is the crystal direction normal to the disc, is usually <100> to minimise the number of surface states which are generated at the interface between the substrate and the layers of oxide which are required for the manufacture of devices. The orientation within the plane of the disc is also important and is indicated by a 'flat' which is ground on the edge of the boule before it is sliced into wafers.

Wafers can be produced by both CZ and FZ with resistivities in the range 0.001 Ωcm to greater than 1000 Ωcm. The bulk material required for device fabrication will often be a few ohm-cms either n-type or p-type but will vary according to which semiconductor process will be used and which device type is to be manufactured. In some circumstances where low resistivities are required, a new layer of silicon of the desired electrical parameters will be grown on the substrate from the gas phase at about 1000°C. This is known as an epitaxial layer and is commonly used in bipolar technology to produce a sharp interface between two regions of widely different resistivity. Care must obviously be taken during the growth to produce sufficiently good crystalline perfection within the layer. In the early days defects known as stacking faults were a well known cause of device failure through high base leakage currents. The whole question of quality of the starting material for device fabrication is becoming more important as smaller dimensions and more sensitive devices are produced.

2.3 LITHOGRAPHY

One of the most important techniques associated with semiconductor device fabrication, lithography is concerned with the definition of circuit details on a mask and the subsequent replication of the mask onto the wafer. A number of different technologies are used. The most frequently used is photolithography which involves the use of visible or ultra-violet light and is capable of defining patterns down to approximately 2μm. For smaller patterns electron-beam lithography has been developed to a commercial stage for dimensions down to about 0.5μm and in the research laboratories x-ray lithography has achieved patterns of only a few hundred Angstroms. A promising new technique using ion-beams is also being investigated for high resolution patterns in the region of 100 Å but this has still to reach commercial exploitation.

The way in which the various techniques are currently used is summarised in Fig. 2.1.

By far the most common route is the purely optical one using photolithography. In this technique a reticle at x 10 final size is produced using an optical pattern generator. The reticle is stepped and repeated while at the same time reducing the size of the image to provide master multi-image masks of the final size circuit geometries. These can either be used directly in a projection mask aligner in which the image of the mask is projected onto the wafer without the two coming into contact or, alternatively, copies of the master mask can be made for use in a contact printer in which the mask and wafer are brought together to form a contact print.

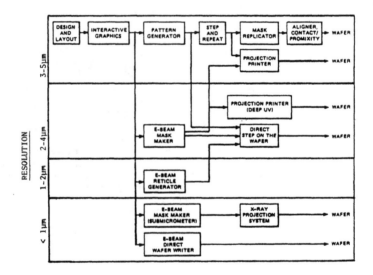

Fig 2.1   Lithographic techniques for device definition

The two techniques are exactly analogous to using an enlarger or making a contact print in normal photography. However, because the pattern definition required is extremely high the equipment used is much more expensive than even the best standard photographic equipment.

For both the initial mask making and the wafer patterning a photosensitive polymer known as photoresist is used, to record the pattern. The photoresist may be positive acting where the material exposed forms the final pattern of the device structure or negative acting when the regions which are not the devices are defined. In the case of mask making a one stage masking process is sufficient. The required pattern is exposed in the photoresist as a series of rectangular images by the pattern generator. The unexposed photoresist is removed chemically and the pattern etched in the thin layer of metal, usually chromium or iron oxide, on the glass plate to form the mask. When the mask is subsequently used to define the patterns on the silicon wafers which will be subjected to high temperature processes, the above procedure is used to define the pattern in photoresist which is etched into an oxide layer on top of the silicon and thus contiguous oxide mask is used as the mask during manufacture. The sequence of events is demonstrated for contact printing in Fig. 2.2.

The wafer is first oxidised and then coated with photoresist. The glass photo mask is placed in contact (correctly aligned, as this process is used many times in the manufacture of devices) and the pattern exposed. The unexposed photoresist is removed using organic solvents and the pattern transferred to the silicon dioxide by etching

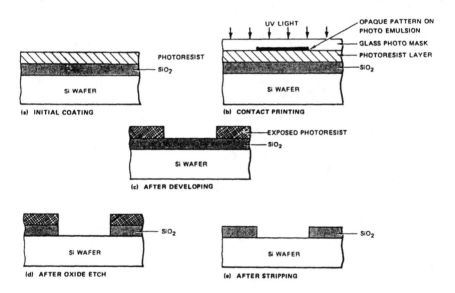

Fig 2.2   Basic photoresist/oxide masking

the oxide through the gap in the photoresist with a buffered solution of HF (hydrofluoric) acid. The exposed photoresist is resistant to this etch and the pattern can be accurately transferred. The unwanted remaining photoresist can then be removed chemically.

The contact printing method gives good definition over small areas but it does damage the masks and introduces mechanical defects into the silicon. Also, if the wafers are not flat the technique suffers from diffraction effects when used over large areas. The more common technique now used is projection printing and alignment where an optical system projects the mask pattern with one-to-one image size onto the wafer. This avoids the above problems but does require good and hence expensive optics. As dimensions reduce it becomes impossible with one-to-one projection over the whole wafer to maintain simultaneously registration and definition. For high resolution work with geometries in the 1-2μm region, projection systems known as wafer-steppers have been introduced in which reticles (single images of the circuit pattern) are reduced 5 or 10 times and stepped and repeated directly onto the wafer with individual device alignment and focus adjustment for each device. This of course is much slower than exposing the wafer with all devices in parallel but is necessary to obtain the placement accuracy for VLSI devices such as 64K RAMS. There is a consequential increase in cost of the equipment for this more accurate technique which is typical of the ever increasing capital investment necessary to keep at the forefront of device manufacture.

Although wafer steppers overcome the problem of replicating small device geometries at least down to 2μm, the capability of producing hundreds of thousands of devices on a single chip produces another difficulty, namely the time taken to produce the reticle which requires several million exposures or 'flashes'. This takes many hours to produce, making the generation of a complete mask set which contains about ten different masks, a long and tedious process. A more rapid technique which is now widely adopted is the use of an electron beam pattern generator in which electron-sensitive resist is exposed by an electron beam. The beam can be moved very much more rapidly than the mechanical stage of an optical pattern generator and hence production times are cut by about an order of magnitude. Note that it is for the rapid production of standard geometries (for reticles) that most e-beam machines are used commercially. The enormous capital cost involved makes the use of such equipment for directly writing patterns on wafers mainly a research activity at present. For very high frequency devices e-beams are now used commercially but the market is small and highly specialised and hence the devices can command a high price.

In order to overcome the serial nature of e-beam pattern generation when applied to direct wafer exposure a refinement of the optical technique has been developed in which a special mask is made with an e-beam machine and used with an x-ray exposure system to replicate, in parallel, the device images. The masks however are extremely fragile and suffer from lack of stability which is not allowing the technique to be widely adopted commercially. The cost of the equipment is, however, in principle relatively modest and high wafer throughput rates should be possible, making it an attractive technique for submicron device geometries.

2.4 OXIDATION

In the previous section the use of silicon dioxide (SiO_2) as a high temperature masking material was briefly mentioned. The production of oxide films and their properties is fundamental to silicon device manufacture and thus will be considered in more detail. The uses to which SiO_2 is put include :

    (i)   Diffusion Mask
    (ii)   MOS Gate Insulation
    (iii) Field Isolation
    (iv) Dopant Source
    (v)   Insulator between Conductor Layers
    (vi) Passivation

Growth of SiO_2 can be achieved most easily by heating up wafers to between 700° - 1250°C in an oxygen ambient. This may be either dry from a gas supply or wet through the introduction of steam into the gas supply. The oxidation process consists of oxygen diffusing through the growing SiO_2 layer and reacting at the silicon surface. In this process the silicon is therefore consumed. The rate of oxidation is determined by many factors but in general is faster from a steam source than dry gas. However, the electrical characteristics depend on the method of growth and hence the correct method must be chosen for a particular application. As a diffusion masking material, the electrical properties are not of concern as it is the relative diffusion rates of the dopant species which are important. For the most common dopants, a mask thickness of about 0.5μm is typically used and hence the faster steam growth is used with a relatively high wafer temperature to grow the mask in about 30 minutes. Such a technique can also be used to form an insulating layer on top of polysilicon interconnection, before either another layer of polysilicon or a metal layer is deposited.

On the other hand where the oxide quality is paramount in determining the device characteristics as in the gate of an MOS transistor, the highest quality is required. Until recently only oxidation from a gas or an evaporating liquid source was used with suitable additives such as chlorine to reduce the free interface states. The oxygen is often mixed with nitrogen as a carrier gas to control the oxidation rate and to reduce the cost, as only a fraction of the available oxygen is used in the reaction. More recently high quality oxides have been produced by "burnt hydrogen" systems in which water is formed in the furnace tube by burning high purity hydrogen in high purity oxygen. Growth rates of oxides with both approaches can be increased by raising the pressure under which the oxidation takes place. This is particularly attractive to increase the rate of dry oxidation at lower temperatures which prevents the unintentional movement of dopants within the wafers. It also reduces the generation of defects within the silicon.

Another approach to oxidation is the deposition of oxide by means of chemical vapour deposition (CVD). A simple reaction is the oxidation of silane:

$$SiH_4 + O_2 \rightleftharpoons SiO_2 + 2H_2$$

With this method, doped oxides can easily be produced

by mixing with the silane the appropriate dopant gas.   Such
oxides act as dopant sources in subsequent processing.   The
temperature can be kept low yet high growth rates can be
achieved by CVD. There are difficulties however, in
minimising the contaminants. Improvements to the basic
technique can be made by reducing the pressure of the
reacting vessel and enhancing the reaction by introducing a
plasma discharge.   Higher uniformity and quality can be
produced with these additions.
       It is worth mentioning that silicon nitride can also
be deposited in a similar way to provide a higher dielectric
constant insulator and this is often used in conjunction
with oxide films in device fabrication.

2.5   EPITAXIAL GROWTH

       Where it is necessary to make devices in heavily-doped
silicon and to avoid high temperature diffusion for long
times, a layer of epitaxially grown silicon is used.   If
silane (SiH_4) is passed over a wafer of silicon at
approximately 1000°C it will decompose and a single crystal
layer of new material, oriented with the substrate, will be
produced.
       Careful control of gas flows and pressures of the
dopant, carrier and reactive gases is required to ensure
good quality and uniform electrical characteristics.   A
typical epitaxial reactor   is shown in Fig. 2.3.

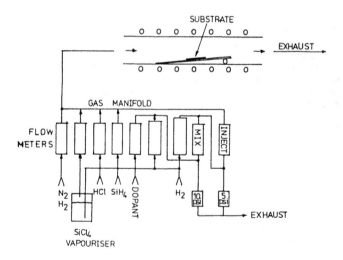

Fig 2.3   Schematic of epitaxial reactor

An important use of epitaxy is in the growth of

silicon on an insulating crystal substrate such as sapphire to produce silicon on sapphire (SOS) technology. Although the substrate is of different crystal structure, there is sufficient lattice matching to allow good quality layers of 0.5 - 1.0 μm to be produced. The attraction of this is obviously that as the sapphire substrate is insulating, individual devices can be isolated from each other by etching through the epitaxial layer to leave separate islands of silicon in which the devices can be built.

2.6 DOPING

The essence of any semiconductor device is the local variation of electrical charges within the material and this is achieved either by allowing impurities to diffuse into a uniform substrate at a high temperature or by implanting the ions directly from a high energy source.

2.6.1 Diffusion

Although diffusion is being replaced by ion implantation (which is more controllable and can produce more abrupt junctions in devices) diffusion is nevertheless less expensive and is still widely used. Most commonly a three-zone furnace is used to provide a uniform temperature zone over the length of a boat load of silicon wafers. The number of wafers varies according to the volume of the plant but in a major facility, boats of 200 wafers will typically be processed as one batch. Again good control, particularly of temperature is essential with a maximum variation of only ±1/2°C in 1000°C being possible. The dopant source materials, boron, phosphorous and arsenic can be in gas, liquid or solid form. The latter two are heated and the vapour mixed with a carrier gas which is passed through the furnace tube. An alternative, which is commonly used for MOS fabrication, is to put solid sources of dopant in the form of oxidised wafers between the silicon wafers in the furnace tube itself.

The method of diffusion is to perform a "predeposition" in which a constant surface concentration of dopant is maintained and follow this by a "drive-in" during which the temperature is increased but no new dopant is provided. The use of a lower temperature of around 800°C for the predeposition reduces the surface damage while the higher temperature of drive-in minimises the time taken to produce the desired diffusion profile.

2.6.2 Ion Implantation

The process of ion implantation consists of accelerating dopant ions in an electric field and injecting them into the silicon. The process is very different from diffusion but ends up producing the same results. In this case the depth of penetration of the dopant ions is controlled by the accelerating voltage and the dose by the flux of ions and the time of exposure to the beam. The advantages are that:

(1)   doping levels are easily controlled from 10^{16} ion/cm^2, equal to the heaviest-diffused predeposition, down to less than 10^{11} ions/cm^2.

(2)   uniformity across a wafer can be readily achieved.

(3)   it is a low temperature process which means that photoresist masking can be used hence eliminating the two stage photoresist/oxide masking necessary with diffusion.

(4)   highly pure species can be selected by mass analysis.

(5)   sideways "diffusion" is eliminated.

The major use of ion implantation has been the accurate control of threshold voltages particularly for low voltage processes such as those used for watches and battery operated calculators.

The main limitation of implantation, other than the high cost of the equipment, is that depths of doping are restricted to about 1μm by the practical accelerating voltage of 200kV. It is more suited therefore to MOS processing than bipolar especially for the buried layers. A further complication is that the process of implantation results in considerable damage to the crystal lattice. This damage has to be removed by annealing the wafers at temperatures between 450°C and 950°C. At the low end of the range the mobility of the electrons is only partially recovered while at the high end complete recrystalisation of the material is effected and bulk values of mobility and carrier lifetimes are fully recovered.

2.7 METALLISATION

High-quality thin-films of metals, normally aluminium which has been slightly doped (3%) with silicon, are required to interconnect devices. These films are produced either by evaporation or cathode sputtering. The former technique is the simpler and is widely used by industry. The metal alloy to be evaporated has to be heated and this can be done in a number of ways. Direct resistance heating is only practicable with resistive materials but is used for refractory metals such as tungsten. For aluminium, it can be evaporated from a resistive heater or by means of r.f. induction heating. Both methods suffer from a tendancy to contaminate the film and the preferred method is by electron bombardment of an aluminium target as source in a localised area. The container of the source can then be kept at a low temperature and does not contribute any contaminants. The evaporation has to take place in a moderately good vacuum ($<10^{-6}$ torr) with the substrates raised to a temperature of a few hundred degrees to ensure the growth of stable films with the correct grain structure.

The alternative cathode sputtering technique involves bombarding a target of the material which is to be coated onto the wafer with an ion beam in the presence of a plasma discharge. The detailed mechanisms involved are complex as the substrate is bombarded by inert gas molecules, electrons, photons and negative ions from the plasma as well as the desired atoms from the target. The major benefit of sputtering arises when a layer of a multi-component material is required. With suitable control of the conditions, such compound layers of defined composition can be produced but they are not widely used in silicon device fabrication. Because of the simple metals required and the faster film growth rate evaporation is preferred for metallisation. Sputtering techniques are, however, being widely adopted in a modified form for pattern definition as an alternative to etching.

As circuit complexities increase, interconnection of the devices becomes a major problem if only one layer of metallisation is available. The use of polysilicon and diffusion for interconnection is not possible for tracks of any length due to the speed restrictions and in the case of gate arrays the structures other than the metallisation are predefined to a standard pattern. Various techniques for multilayer metallisation are therefore being developed and particularly in the gate array field two layer schemes are common. The first layer of interconnect, usually aluminium is deposited and patterned in the normal way. This is covered by an insulating layer of polyimide which is "spun" onto the wafer in the same way as photoresist. Contact points to the underlying metal are made with an oxygen plasma dry etching process before the final aluminium layer is sputtered and patterned. This is discussed below.

2.8 ETCHING

This is one of the areas where there have been rapid recent developments to enable finer geometric dimensions to be produced but has always been important as the geometry of semiconductor devices are closely related to their electrical characteristics. Having defined a pattern in photoresist as described earlier it is necessary to etch the pattern in the silicon with the minimum of dimensional change. Ideally the etch should produce vertical sidewalls in the etched pattern but in practice many effects combine to give a degree of undercutting of the photoresist. Using standard wet etching techniques, the etchant has to be selected carefully to ensure the appropriate degree of isotropy is obtained. In the case of oxide etching for instance, a solution of HF buffered with ammonium bifluoride is usually used to give a minimum of undercutting. The rates of etching depend on the method of oxide growth, and the impurity doping as well as the acid mixture. Because of its crystalline nature the situation is more complicated with silicon and much study has been given to the development of etches which etch different crystal planes at different rates. For instance water-ethylene diamine pyrocatechol (EDA) etches the <100>, <110> and <111> planes in the ratios 50:30:3 μm/hr respectively at 100°C. This facility can be used to good effect to produce either V-groove structures as in V-MOS or parallel sided cuts necessary for fine lines as shown in Fig. 2.4.

More difficult to etch is silicon nitride which requires treatment with hot orthophosphoric acid at about 165°C. This is too high a temperature for photoresist so that a thin oxide mask which is unaffected by the nitride etch is used. A similar technique has to be used for polysilicon which also requires an intermediate oxide mask. More complicated still is the etching of aluminium which

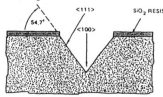

(a) THROUGH A PINHOLE <100> ORIENTED Si

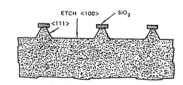

(b) THROUGH A SQUARE WINDOW FRAME PATTERN <100> ORIENTED Si

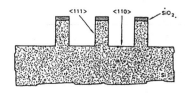

(c) THROUGH A SQUARE WINDOW FRAME PATTERN <111> ORIENTED Si

Fig 2.4   Selective etching of silicon

requires a carefully controlled mixture of orthophosphoric, nitric and acetic acids. A slightly raised temperature (50°C) is used which allows the use of photoresist masking but care must be taken as the photoresist is attacked by the nitric acid.

Although wet etch techniques have proved successful for geometries down to a few microns, they are difficult to control and in general it is not possible to produce suitable anisotropic etches which are necessary to cut very narrow channels through relatively thick masking materials. As a result, dry etching techniques have recently been developed. The main techniques used are as follows:

> Plasma etching
> Sputter etching
> Ion Milling
> Reative Ion etching

In plasma etching, wafers are subjected to reactive plasma which has been generated by a r.f. discharge at low pressure. In order to obtain uniform etching the short lived ions are screened by means of a perforated plate around the wafers which is electrically grounded. The long lived species penetrate the plate and attack the wafers which are mounted as in a diffusion furnace about 5mm apart in a jig. The temperature of the wafers can be kept low so that the photoresist masking is adequate. Most materials of interest can be etched using fluorine and/or chlorine based gases. Indeed one of the difficulties is producing sufficient selectivity which is the basis of pattern production.

In sputter etching the basic sputtering process is used to remove material from the required regions. Relatively high pressures, about 50mTorr are used but this has the effect of introducing radiation damage from the back-scattered particles which are of sufficient energy to affect MOS devices.

Ion milling is a similar process carried out at lower pressure. (<0.1 mTorr) This overcomes the radiation damage problem but as with sputter etching the material removal is almost completely non-selective with the rate of removal of resist approximating to that of the desired material.

Finally, reactive ion etching is simply sputter etching using a reactive gas at a pressure of approximately 20 mTorr. Appropriate selection of the reactive species improves the selectivity of device etching materials with respect to photoresist removal. For example, using etchant CHF_3 an etch ratio of 6:1 can be obtained for SiO_2: photoresist.

In order to produce etched geometries of less than $3\mu m$, where anistropic etching is essential for dimensional control, it is apparent that difficulties of dry etching are still to be overcome. Production processes inevitably include a variety of plasma and reactive ion etching techniques. This will be determined by the sequence of materials to be etched and the relative sensitivity of each process step to radiation damage and cross contamination.

2.9 MANUFACTURING PROCESS

The use of these techniques to produce a complete integrated circuit is summarised in Fig 2.5 which gives the various manufacturing stages for both bipolar and MOS processes. As can be seen, the number of steps required for the bipolar device is greater and this, together with the fact that the devices cannot be packed so closely, means that the most popular technique for large and very large scale integration is MOS. The basic MOS processing sequence is illustrated in Fig 2.6 but behind this simplistic view are about fifty steps.

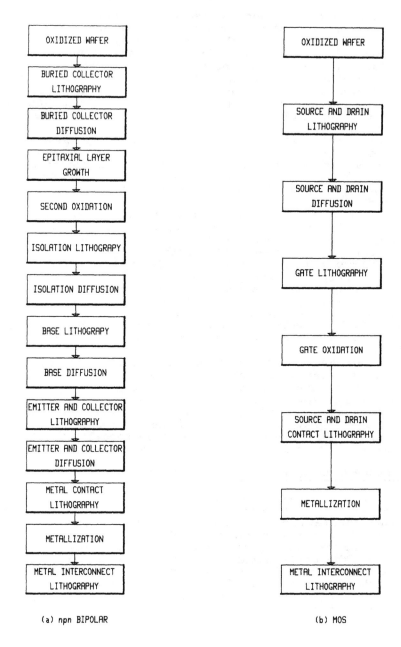

(a) npn BIPOLAR

(b) MOS

Fig 2.5   Fabrication processing for (a) bipolar
(b) MOS transistors

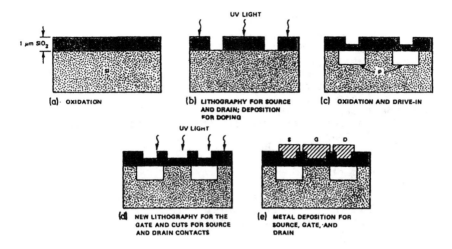

Fig 2.6   Simple metal gate MOS process sequence

The schedule of events for a straightforward n-channel silicon gate process is as follows:

N-CHANNEL SILICON GATE PROCESS

1. Starting Material :   14-20 ohm.cm.<100> P-type, 3in.Dia.
2. Initial Clean
        Removal of organic material and silicon dioxide in mineral acids.
3. Initial Oxide
        500 Angstroms thermal oxide.
4. Silicon Nitride Deposition
        500 Angstrom low pressure chemical vapour deposition.
5. Masking Oxide
        Thermal oxidation of nitride surface for selective resistance to nitride etch.
6. 1st Photo (Positive Resist) LAYER 1
        Masks off active device areas and diffusion lines with photoresist.
7. Boron Implant
        Implants through nitride and oxide those field regions not masked by resist to increase the final field inversion voltage.
8. Oxide Etch
        Removes surface oxide from nitride layer in field regions not masked by resist.
9. Resist Strip
        Resist dissolved in fuming nitric acid.
10. Silicon Nitride Etch
        Nitride removed in Orthophosphoric Acid at 165°C from field regions not masked by surface oxide.
11. Field Oxide
        Thermal oxidation in steam. 1 micron SiO_2 grows

selectively in areas not masked by nitride.

12. Oxide Etch

Short etch removes surface oxide from nitride blocks without significant removal of field oxide.

13. Silicon Nitride Etch

Nitride removed from active areas in Orthophosphoric Acid at 165°C without etching field oxide.

14. Oxide Etch

Removes remains of initial oxide from active areas.

15. 2nd Photo (Positive Resist)   LAYER 2

Opens windows in resist over depletion gate regions.

16. Arsenic Impant

Inverts the surface of depletion gate regions with arsenic implant dose centred at a depth of approximately 0.9 microns.

17. Resist Strip

18. Gate Oxide

800 Angstrom chlorinated thermal oxide grown in gate regions.

19. Boron Implant

Both enhancement and depletion thresholds finally set with low energy implant in all active areas.

20. Resist Strip

Removes contaminating organics associated with implanter vacuum system

21. Anneal

Anneal in nitrogen activates implant and reduces gate oxide/silicon interface states to a minimum.

22. 3rd Photo (Positive Resist)   LAYER 3

Defines direct contacts from polysilicon to diffused areas.

24. Oxide Etch

Etches direct contact areas back to bare silicon in diffusion regions.

25. Resist Strip

26. Polysilicon Deposition

3500 Angstroms of polysilicon deposited by low pressure chemical vapour deposition.

27. Poly Oxide

Thermal oxidation of polysilicon to act as an etch mask.

28. 4th Photo (Positive Resist)   LAYER 4

Defines polysilicon gates and interconnects.

29. Oxide Etch

Stensils the polysilicon lines in masking oxide.

30. PolySilicon Etch

Etches polysilicon as defined by the masking oxide and photoresist.

31. Oxide Etch

Removes both the poly masking oxide and gate oxide exposed by polysilicon etch.

32. Phosphorus Deposition

Deposition dopes both the polysilicon and the

source/drain regions opened up in the previous stage.

33. Phosphorus Deglaze
    Short oxide etch to remove excess deposited phosphorus glass and improve uniformity of final sheet resistance.

34. Poly Oxide
    2000 Angstrom thermal oxide assists in reducing metal/poly and metal/diffusion shorts caused by pinholes in the following pyrolytic oxidation step.

35. Reflow Pyro Deposition
    7500 Angstroms of heavily phosphorus doped silicon dioxide deposited by pyrolytic decomposition at low temperature.

36. First Reflow
    High temperature oxygen anneal, melting the doped oxide to produce a smooth coverage of sharp poly silicon edges.

37. Densification of Reflow Pyro
    Short wet oxidation leaches phosphorus from the surface of the doped oxide to promote adhesion of photo resist.

38. 5th Photo (Positive Resist)   LAYER 6
    Defines contacts between metal and polysilicon or diffusion.

39. Reflow Etch
    Oxide etched completely from contact areas.

40. Resist Strip

41. Second Reflow
    High temperature anneal in nitrogen to round off etched contact edges for improved metal coverage.

42. Aluminium Evaporation (Si Gate)
    1.5 microns 99.995% pure aluminium electron beam evaporated at 200°C.

43. 6th Photo (Positive Resist)   LAYER 7
    Defines aluminium interconnection pattern.

44. Aluminium Etch
    Pattern etched in Aluminium Etch at 50°C.

45. Resist Strip

46. Sinter
    Low temperature anneal at below the aluminium/silicon eutectic in a nitrogen/hydrogen atmosphere to tack metal to contacts and eliminate electron beam radiation damage.

47. Overlay Pyro Deposition
    7500 Angstrom deposit of lightly phosphorus doped pyrolytic oxide for mechanical protection of aluminium pattern and reduced sensitivity of devices to ionic contamination.

48. 7th Photo (Positive Resist)   LAYER 8
    Defines windows in overlay pyro oxide over aluminium bonding pads and test points.

49. Overlay Pyro Etch
    Oxide removed from bond pad areas.

50. Resist Strip

Electrical Parameters

Parameter	Process Value	Conditions (all with back bias of -2.5 volts at 21°C)
Enhancement Threshold voltage	1V	Zero drain current Channel length = 6μm
Depletion Threshold	-4V	
Back bias constant	(enhancement) 0.3V	Channel length = 6μm - increase by 25% for channel length = 8μm and by 30% for channels greater than 15μm.
Field inversion voltage	> 20V	
Enhancement Gain	30 μA/V^2	Zero gate voltage Unity aspect ratio
Depletion Gain	24 μA/V^2	Same as for enhanceme...
Diffusion Resistance	7-20 Ω/sq	
Poly Resistance	20-70 Ω/sq	
Metal Resistance	0.03 to 0.04Ω/sq	
Metal to Substrate Capacitance (over field oxide)	2.5$\pm$0.6 x 10^{-5}pF/μm^2	V_{metal} = 0V
Metal to Poly Capacitance	5$\pm$1.6 x 10^{-5}pF/μm^2	V_{poly} = 0V
Poly to Channel Capacitance (gate oxide capacitance)	3.5$\pm$0.3 x 10^{-4}pF/μm^2	
Metal over Diffusion Capacitance	4.3$\pm$0.3 x 10^{-5}pF/μm^2	
Poly to Substrate Capacitance	5$\pm$0.8 x 10^{-5}pF/μm^2	
Sideways Diffusion	1.25μm	
Diffusion Depth	1.5$\pm$0.2μm	
Max. Junction Temp	115°C	

Normal Operational Voltages:

V_{DD}	5$\pm$0.5 volts
V_{SS}	0 volts
V_{BS}	-2.5 volts

Table 2.1   Electrical design rules for a 6 μm n-channel silicon gate process

In a state-of-the-art memory process there would be many additional stages including the use of more than one layer of polysilicon and possibly the use of refractory metals and implanted resistors. Also for a complementary MOS process (CMOS) there is the added complication of building both n-channel and p-channel devices sequentially in the same substrate. The number of mask layers required for such a process will be 12-15 and the number of process steps in excess of one hundred!

2.10   DESIGN RULES

An important consideration which is closely associated with wafer fabrication is the specification of the process in terms which are understandable by designers. This is done by means of a set of "design rules" which define the important electrical and minimum physical parameters which can be guaranteed by the process engineers. The existence of a well-defined set of rules is crucial to the successful production of integrated circuits where the two major stages of manufacture, design and fabrication, are distinctly separate activities. Even so it is often the interface which causes the difficulties. As part of the design rules such information as which mask layer refers to which fabrication step must be defined as will position and style of alignment marks, test structures, numbering sequences and details of scribe channels between devices. Although in principle such information could be standardised it tends to vary slightly from manufacturer to manufacturer.

Rules, which are specific to a particular process, however, relate to the electrical characteristics such as transistor thresholds and gains, capacitance between various layers and resistances. An example of electrical rules for an N-Channel Silicon gate process running at the University of Edinburgh are given in table 2.1.

Associated with these parameters and equally important are the minimum geometrical rules. These define all critical dimensions between structures within the devices and are quite lengthy. An extract from the geometrical design rules for the same process as described above is given in table 2.2 for the polysilicon areas.

Detail	Dimension (μm)
Width of track as interconnect	6
Width of track as gate	6
Separation between polysilicon tracks	6
Overlap on thick oxide at channel side	4.5
Overlap round contact windows	3
Separation from unrelated diffusion or parallel related diffusion	3

TABLE 2.2   Extract of Geometrical Design Rules relating to polysilicon for a 6μm N-channel Enhanced Silicon Gate Process.

2.11 NEW TECHNIQUES

One of the major driving forces of the technology is the desire to produce more and more complex (VLSI) circuits or to produce higher frequency devices. The American military programme in the latter area is known as VHSIC (Very High Speed Integrated Circuits). No one aspect of device fabrication will yield all the improvements sought and there are continual developments to reduce line widths, capacitances and power consumption by whatever means is available. If one looks at the fundamental physical limitations of devices, which are related to quantum and statistical physics, it is estimated that about $0.25\mu m$ will be the fundamental limit for MOS device channel length. Likewise an oxide gate thickness of about $10^{-3}\mu m$ is considered the minimum before quantum mechanical tunnelling effects give rise to a significant device current. Similar limits can be put on the minimum dimensions for bipolar transistors.

In the lithography process electron beams and x-ray exposure are easily capable of producing dimensions in the range 100 - 250Å. X-rays appear to give the best long term prospects but improved mask stability is required. Special sources of X-rays using accelerated electrons have been developed for X-ray lithography with high yield of characteristic radiation about 10Å wavelength. Hot plasma generated by high power lasers or electrical discharge have also been tried and look interesting while considerable experimental work has been done using synchrotron radiation. The advantage of this latter source is that the radiation available has a very wide spectrum and an appropriate wavelength can be selected from the beam by Bragg scattering.

For doping, wider use of ion-implantation is desirable to achieve lower temperature processing and hence better defined junctions, which are necessary for small geometry devices. As previously described the damage introduced by the process has to be annealed out by annealing the wafer, albeit for a shorter time and perhaps at a lower temperature, than for diffusion. However, as deeper, more energetic implants are carried out, the benefits become outweighed by the problems introduced by the higher temperatures required for wafer annealing. To overcome this, localised annealing using a high powered laser beam is being adopted. Very tight geometric control of the recrystallisation zone is possible and good control over the dopant diffusion depth can be achieved by varying the power and duration of the laser pulse. Also since laser annealing is an extremely rapid process it is not necessary to use a special inert atmosphere to prevent oxidation or contamination during the annealing procedure. For silicon or germanium, implanted with Arsenic, Boron or Phosphorus, either a Q switched Neodymium Yag or a ruby laser can be used with a pulse length of 10 -100 ns and power of 0.5 - 10J/cm^2. The time during which the exposed material is molten is only a few hundred nanoseconds so rapid scanning of the complete circuit is possible.

The use of laser annealing to produce single crystal material from polycrystalline material has also been reported. The importance of this work is that it opens the way to building three dimensional structures if devices can be made in recrystallised polysilicon on top of an insulating layer such as oxide. For this some "seeding" is of course necessary but this can be achieved by suitably patterning the oxide with a fine enough grid.

The major area for future development which currently limits the sizes of devices which can be produced is the etching process. The industry is at present at the stage of transferring from wet to dry etching and many elegant demonstrations have been given of very fine line production using sufficiently thick masks to enable subsequent processing to take place effectively. These rely on the use of nonisotropic etching which in the plasma and reactive ion etching field is at an early stage of development. As the techniques become better defined and industry adopts them, progress towards the fundamental limitation will be rapid. It is difficult to believe, however, that no further improvements will be made beyond those already mentioned. Perhaps new physical effects will be discovered in the existing materials which will allow higher performance designs to be produced. More likely is the possibility that new materials will be adopted, perhaps in the domain of organic molecules. Certainly if nanometric dimensions are to be a reality a different level of understanding of materials and surfaces will be required.

Chapter 3

Review of bipolar
and MOS technologies

A.D. Milne

3.1 INTRODUCTION

The rate of technological change in electronics has
been quite extraordinary over the past twenty years and
while there is little sign that new developments will cease
in the foreseeable future, some semblance of order or even
stability is beginning to appear, at least in the field of
silicon integrated circuits, where the technology is
reaching maturity. Other devices such as those based on
gallium arsenide (GaAs) are only at the start of their
evolution in everyday commercial use and no doubt we will
experience a hectic period of development while that
technology matures. The same is true of the more abstruse
technologies such as those using macroscopic quantum
effects, the so-called Josephson Junctions, which are a most
interesting, if rather specialised, area of development. In
this chapter the focus will be on the mainstream
technologies based on silicon which, to date, constitutes,
for all practical purposes, the only available material for
the construction of large or very large scale integrated
(VLSI) circuits.

The range of technologies based on silicon is however
quite varied. Different approaches have been developed by
different companies for particular applications and until
comparatively recently, there were almost as many
fabrication technologies as there were semiconductor
companies. Commercial pressures together with the maturing
of the technologies mentioned above are, however, leading to
some rationalisation. Companies are beginning to undertake
joint developments and the need for second sources of large
volume products has led to considerable cross licensing of
technology. As a result there have emerged a number of
"industry standard" or preferred technologies. Of course,
these basic styles of circuit manufacture continue to be
developed, primarily through the reduction of the physical
dimensions of the device features. It is this process,
known as device "scaling", which is the principle reason
behind the increase in complexity which is possible in a
single integrated circuit. The renowned Moore's law (1),

illustrated in Fig 3.1, quantifies the effect as
approximately a doubling of components per chip each year.
An important consequence of this, which of course is the
major driving force in the first place, is the corresponding
reduction in the cost per function. Expressing this in
another highly dramatic way, any integrated circuit costs
$3, no matter how much circuitry it contains.

    With increased automation and the use of electron
beams or X-rays for direct writing on the wafer,
considerable further development is possible. By the end of
the decade it is predicted that sub-micron feature sizes
will be in production which will allow memories considerably
in excess of 1 million bits of storage to be produced as
single chips. Such technological advances will also apply
to other types of circuits such as microprocessors, single-
chip microcomputers and, of course potentially to gate
arrays.

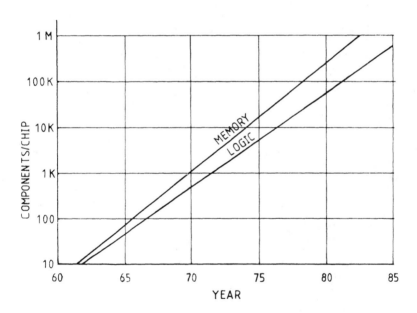

Fig 3.1   Growth of component density/chip

3.2   SILICON TECHNOLOGY

    The two main branches within silicon technology are
bipolar and Metal Oxide Semiconductor (MOS) technologies
(Millman (2)). Within these broad categories there are many
subsections which have particular characteristics, as shown
in Fig 3.2.

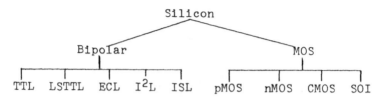

TTL - Transistor-Transistor Logic   pMOS - p-channel MOS
LSTTL - Low Power Schottky TTL   nMOS - n-channel MOS
ECL - Emitter Coupled Logic   CMOS - Complementary MOS
I²L - Integrated Injection Logic   SOI - Silicon on Insulate
ISL - Integrated Schottky Logic

Fig 3.2 Silicon technologies

Bipolar technology, conceptually more advanced than MOS, was the first to yield successful discrete devices and later the first integrated circuits. Although complex and requiring a larger number of process steps, all the active semiconductor junctions in a bipolar device are within the bulk of the silicon and are therefore immune from surface contaminants. Various families of logic (eg RTL, DTL, TTL, ECL) have been produced over the years using bipolar technology (Garrett (3)). At the present time the most common variants for logic are TTL, the low power Schottky version LS-TTL, and ECL with integrated injection logic (I²L) (Hart and Slob (4)) and integrated Schottky logic (ISL) (Lohstroh (5)) used for both gate arrays and LSI custom devices. The basic logic circuit technologies are also used for gate arrays although because of their relatively high power dissipation per gate, the size of the arrays is limited. Another important gate array technology is the Collector Diffusion Isolation process (CDI) (Murphy (6)) used by Ferranti but not otherwise widely adopted.

MOS technology was initially developed to meet the needs of lower performance circuits with higher levels of integration i.e more devices per integrated circuit. The simpler processing and fewer mask layers required has allowed larger circuits to be produced than with bipolar techniques while achieving an adequate yield for commercial success. Early devices made use of the simple pMOS technology which does not require an implantation to adjust threshold voltages. With the availability of ion implantation the improved performance of nMOS was welcomed and more recently the combined technology of complementary MOS (CMOS), which offers lower average power consumption per gate, has begun to increase in popularity. As power consumption is perhaps the key factor in the design of VLSI circuits it is widely accepted that CMOS will be the dominant technology of the future.

Within the basic CMOS design technique a number of variations are possible. Improvements in speed over the basic bulk silicon CMOS can be made by constructing devices on an insulating substrate - Silicon on Insulator (SoI).

The most common insulator is sapphire (hence silicon on sapphire (SOS)) in which the sapphire forms the substrate for the devices which are built in an epitaxial silicon layer (Forbes (7)). Apart from the technical difficulties of manufacture, the cost of the sapphire substrates limits the widespread use of this technology in commercial applications but for military purposes the enhanced speed and hardness to radiation is a strong attraction. More recent alternatives however are to build devices in recrystallised silicon on silicon dioxide insulator. This technology retains the bulk silicon substrate and hence can be cheaper than the SOS approach but requires new techniques such as laser recrystallisation.

3.3 BIPOLAR TECHNOLOGIES

3.3.1 Device Structure and Isolation Techniques

A simplified cross section of a traditional bipolar device is shown in Fig 3.3a. The device, in this case an npn transistor, is constructed vertically with a buried n^+ layer under the collector region to reduce the series resistance to the collector contact and to prevent parasitic transistors being formed with the substrate. Individual devices are junction-isolated by surrounding them with deep p^+ diffusions. Not only are these diffusions expensive in manufacturing time, they also consume valuable silicon area which could otherwise be used for active devices. Methods of improving bipolar circuits have concentrated on improved methods of isolation and the reduction in the thickness of the base region, known as the base width. The most common form of isolation used today is

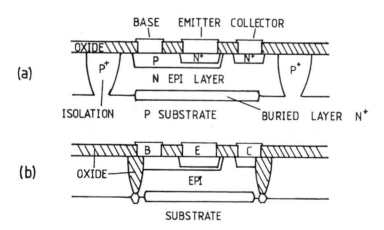

Fig 3.3   Cross section of bi-polar processes
    (a)   Conventional (junction-isolated)
    (b)   Mosaic (oxide-isolated)

oxide isolation, as in the so-called LOCOS or ISOPLANAR processes, in which devices are separated from each other by thermally grown silicon dioxide which penetrates the epitaxial layer (Davidsohn and Lee (8)). Recent developments of this technique involve deep U-shaped grooves which are first oxidised and then filled with polysilicon. As the oxide is inert it can be used to define the edges of the emitter and collector regions as in the self-aligned MOSAIC process of Motorola (Fig 3.3b).

An alternative approach, which is used by Ferranti in their Collector Diffusion Isolation (CDI) process, is to achieve isolation by means of the diffused collector contacts which penetrate through the epitaxial layer (in this case P-type) and join the buried n^+ layer. This obviously has advantages in terms of the sizes of the devices and hence relatively high speeds can be achieved. This results from a reduction in the parasitic capacitance associated with the devices which enables them to operate at higher speeds.

3.3.2 High-Performance Bipolar Technologies

The performance of a bipolar process depends on the width of the base region, the isolation between devices and the alignment of the device elements. The need for increased speed has led to the use of predeposition implantations, thinner epitaxial regions, and polysilicon and oxide isolation to allow self-aligned structures to be easily fabricated. A range of new processes incorporating these features have been developed by Japanese companies, for example NEC's Polysilicon Self Aligned (PSA) technology and Oki Electric Industry Co.'s Base Emitter Self-Aligned Technology (BEST) (Capece (9)). More recently from the Musashino Laboratory of NTT is a super self-aligned bipolar technology (SST1) with a sub-micron base width and a gate delay of only 63 ps. Although this is still at a development stage the technology has been successfully used to produce fast static RAMs - a 256 x 4 bit device with an access time of only 2.7ns. It is predicted also that with device scaling, 1ns access times will be achieved.

American companies such as Motorola have made similar developments for their high-speed MECL devices (MOSAIC process) and Advanced Micro Devices have used their proprietry IMOX process to build a full 16-bit high performance microprocessor (AM29116). The IMOX process is a fully implanted process with minimum feature sizes of 3μm in which densely packed low power oxide-isolated structures are built. The AM29116 utilises ECL techniques for its internal workings to increase the packing density and TTL input and output circuitry for systems compatibility. The result is a most advanced bipolar chip whose size is greater than 4.24 x 4.24 mm with an equivalent gate count in excess of 2500.

An interesting development from IBM has been their use of LS-TTL techniques with small voltage swings to achieve gates with average logic delays of 1.15ns at a power dissipation of 1 milliwatt. Their analysis showed that the power-delay product of 1.15 pico Joules achieved with TTL could not be equalled by ECL which gave approximately 3.2pJ for the same circuit speed. The circuits used for the analysis were 4.5mm square gate arrays containing 704 TTL circuits and 60 emitter follower driver circuits. From these results, they predict that VLSI circuits with greater than 10,000 gates could be built using this technology. The Japanese on the other hand have been pushing towards higher speeds using deep U-shaped grooves and have achieved ECL gate delays of only 170 ps.

3.3.3   Integrated Injection Logic

In spite of the direct attack on large scale integration of bipolar circuits described in the previous section, the more common approach for achieving LSI with bipolar techniques is to use Integrated Injection Logic (I^{2}L) (or Merged Transistor Logic (MTL) as it is sometimes called). This technology was developed as an optimised structure from the layout point of view while retaining the capability of low power operation. By using multicollector transistors operating in an inverted common collector mode (i.e. with common emitters), and with an integrated lateral

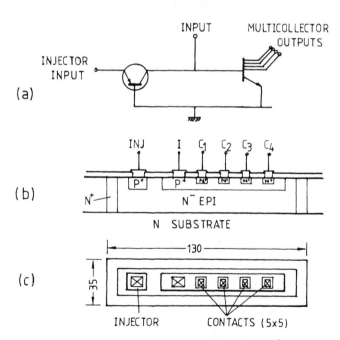

Fig 3.4   Integrated injection logic I^{2}L
          (a) Circuit
          (b) Section through gate
          (c) Plan of layout (μms)

pnp transistor as a current source, an extremely compact merged layout can be realised. Another useful feature is that the technology can operate either with high voltage swings to interface to other technologies and provide good noise immunity or with very low voltages and current and hence power consumption when speed-power products of 0.1 pJ per gate can be obtained.

Also it is straightforward, by means of extra masking layers, to include standard bipolar transistors so that high performance ECL devices can be combined with the low-power, high-density capability of I^2L to provide the potential for high performance VLSI circuits. In the commercial field I^2L has been used by Texas Instruments for their complex 16-bit microprocessor SBP9900.

More recent developments of the I^2L process known as Integrated Schottky Logic (ISL) and Schottky Transistor Logic (STL) by Texas Instruments, Philips Research and Fairchild give even better performance than I^2L with gate delays approaching 1ns and a power consumption below 10nW/gate. These technologies make use of two layers of metallisation with a combination of lateral and vertical transistors in addition to the normal I^2L multicollector structure. With this process very dense gate arrays (120-180 gates/mm^2) with delays of down to 2.7 nS are possible.

Recent announcements by Hitachi and Honeywell indicate a move to higher speeds with the Japanese company achieving gate delays of 290pS using their SICOS (Sidewall Base Contact Structure process while Honeywell has reported devices with delays of 0.6ns at 100 μA collector currents with a 1.25 μm oxide isolated process.

3.4 MOS TECHNOLOGIES

3.4.1 MOS Device Structures

The MOS or field effect transistor differs from the bipolar structure in that it is a lateral device whose dominant parameter is the length of the gate (Fig 3.5). In the early days of the technology the channel, which would be several tens of microns in length, would be controlled by a metal electrode which was usually aluminium (metal gate technology). Both p-channel and n-channel variants were developed for different applications with the latter becoming the industry standard during the early 70's for integrated digital logic circuits. The use of polysilicon, initially introduced to produce higher speed self aligned transistor structures has more recently (late 70's) become the preferred industry process and is used for the bulk of the memory and microprocessor products.

This dominance is now being challenged by the emergence of advanced CMOS processes which compete for speed with nMOS and also offer near zero standby power consumption. As circuits tend towards VLSI the number of devices which can be integrated is determined not so much by the size of the individual devices as by their power dissipation. Hence CMOS technology is preferred to nMOS for

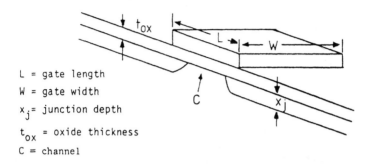

L = gate length

W = gate width

x_j = junction depth

t_{ox} = oxide thickness

C = channel

Fig 3.5   Parameters of device scaling

circuits in which all the devices are not operational at any one time. However, CMOS technology requires both n and p channel devices and is therefore more complex, so until recent developments in isolation techniques become available the area required by each gate was considerably larger than the nMOS equivalent. The rate at which CMOS technology is now being adopted, however, is an indication of its perceived potential. Most semiconductor companies have a CMOS capability and many are now offering improved second generation CMOS products developed from their existing nMOS designs.

3.4.2 MOS Device Scaling

Apart from the change to n-channel devices, the key to achieving higher speed MOS circuits has been the shrinking of the physical dimensions of the transistor (Fig 3.5). The benefits of device scaling are not only to increase speed but also to increase yield by reducing the size of the chips. There also appear to be advantages in reducing the sensitivity of devices to gamma radiation. Almost all companies have introduced products over the last few years which take advantage of reduced geometries. Most familiar will be the HMOS process of Intel and Motorola which itself has been recently replaced by HMOS II, and HMOS III (further reductions), but other new names such as S-MOS of Texas Instruments or X-MOS of National Semiconductors all relate to geometrically scaled down MOS processes. The effects of scaling can be appreciated by reference to Fig 3.5, which gives the relevant dimensions involved. To first order, the characteristics of the MOS transistor remain unchanged if all the dimensions of the device are scaled equally by a factor (1/K) and the doping level of the substrate increased by the same factor (1/K) to maintain the existing threshold voltage and to avoid lowering the punch-through voltage (Table 3.1).

Device Parameter	Scaling Factor	
	Theoretical	Actual
Dimension L, W, t_{ox}	$1/K$	$1/K$
Doping Concentration	K	K
Voltage	$1/K$	1
Propagation Delay	$1/K$	$1/K$ (>1)
Power Dissipation	$1/K^2$	<1

TABLE 3.1   Effects of scaling

This implies that the supply voltage has also to be reduced by $(1/K)$ which, from a systems compatibility point of view is generally unacceptable. In order to maintain the 5v supply, doping concentrations are further increased and, because of the higher gain of the devices, the width W is reduced by an amount slightly greater than K. The important consequences of these modifications are that, apart from the obvious reduction in device area, the power dissipation and power-delay product are both reduced.   From a simple model the reduction should be by the square and cube of the scale factor respectively, but second order effects together with the retention of a 5v supply mean this is not realised and an approximately linear reduction is observed.

There are, however, limits to scaling if the 5v supply level is retained.   Projecting gate lengths down to 1/2 µm, it would appear from geometric considerations feasible to put about 800,000 gates on a chip 6 x 6 mm.   If this were done however, power dissipation would be of the order of 1kW! Assumming a practical limit of say 1W dissipation, the number of gates integrated would be limited to approximately 1000.   On the other hand, scaling the supply voltage in proportion to the geometric gate length reduces the power dissipation per gate and the original 800,000 gates can be readily   integrated within the restriction of 1W overall dissipation.   The difficulties introduced with this approach are that the reduced threshold voltage is more difficult to control and there is a restricted noise margin above the basic kT noise of any semiconductor device.

Typical of the first generation scaled processes, HMOS reduced the gate length of devices from the 6 µm of the standard nMOS process to 3µm.   Further scaling to 2µm was achieved with the second generation HMOS II and more recently to 1.5µm in the latest HMOS III process have given both increased speed and reduced power consumption (Table 3.2). Note that the 5V power supply has been retained.

It is important to remember that not only are devices built with the latest technology superior in technical performance, they are also physically considerably smaller hence they have increased yield in manufacture.   For instance, one of the Intel 8-bit microprocessors was reduced in size by 60% going from HMOS I to HMOS III.   This has

	Std. nMOS	HMOS	HMOS II	HMOS III
Gate Length μm	6	3.5	2	1.5
Junction depth μm	2.0	0.8	0.8	<0.3
Gate Oxide Thickness Å	1200	700	400	250
Power Supply V	5	5	5	5
Minimum Gate Delays nS	4	1	0.4	0.2
Speed Power Product pJ	2	1	0.5	0.25

TABLE 3.2 – Comparison of parameters of scaled MOS devices

important economic effects which it will not be possible to discuss here.

With an improved understanding of the device physics, it has been possible to design these small dimension transistors but fundamental problems have been encountered in reducing the resistance of the polysilicon interconnections between devices. Unless this can be done, the RC time constant of the interconnect is unchanged by scaling which seriously limits the overall chip performance. The solution being adopted by the industry is to replace the polysilicon for both interconnections and gate electrodes by high temperature metals and metal silicides which have resistances two orders of magnitude lower than that of highly doped polysilicon. Although platinum silicides are widely used in bipolar technology their application to MOS devices has proved more difficult and it appears that further development is necessary before the technique is used as a production process. Results from Hewlett-Packard, however, who have successfully produced a 32bit CPU chip with 450,000 transistors using two layers of Tungsten with gate lengths of only 1.5μm, demonstrate the possibilities. In this chip, which is used commercially in their latest HP9000 computer, the main clock frequency is 18MHz and specially designed logic structures give 32b x 32b multiplication in 1.8μs and a 64b by 32b divide can be accomplished in 3.5μs.

A further indicator of the technology is seen in the recent announcement of 256K dynamic RAMs. These are exclusively nMOS designs but most use some form of hard metal for gate electrodes and interconnections, as shown in Table 3.3.

Company	Access Time	Chip Size	Technology
OKI	100 ns	45 mm^2	2 poly nMOS
NEC	100 ns	42.8 mm^2	Molybdenum gate nMOS
Hitachi	150 ns	46.6 mm^2	Polycide gate nMOS (2μm)
Motorola	100 ns	46.5 mm^2	HMOS III (2μm)
Bell Labs	105 ns	54.2 mm^2	Tantalum Silicide Interconnect (2.3μm)

TABLE 3.3 Comparison of 256K DRAM prototypes.

An analysis of the underlying physical limitations suggests that circuits using MOS transistors and polysilion for high value resistors can be scaled by a further factor of approximately 10 (Table 3.4). Allowing for a continuation in the growth of die size it is therefore predicted that chips containing 10^8 transistors could be built by the end of the century!

	Parameter Limit
Channel Length L (μm)	0.22
Supply Voltage (V)	0.36
Oxide Thickness (Å)	50
Channel Doping (cm^{-3})	1.6×10^{17}
Power–Delay product (fJ) (of Inverter)	0.02
Gate Delay (pS)	30

TABLE 3.4 Limits for NOR gate logic

3.4.3 CMOS Technology

The major breakthrough of the past few years has been the application of all the nMOS process developments to CMOS which has transformed what was a rather pedestrian technology into the most promising candidate for VLSI circuits. Traditional CMOS technology uses an n-type wafer with a deep diffused p-type well in which the individual nMOS transistors are built (Fig. 3.6). If all gates are

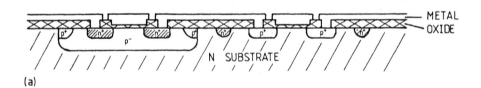

(a)

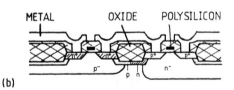

(b)

Fig 3.6 A comparison of CMOS processes
    (a) Metal gate - early 1970s
    (b) Modern silicon gate twin well to the same scale

constructed with pairs of individually isolated devices, large amounts of silicon area are used, but once it is recognised that pMOS devices are only required where low standby power is desired and occasionally with whole blocks of high speed nMOS logic, or for instance for input/output control and analogue functions, high density circuits can be built.

In order to implement such circuits large p-type "ubiquitous" wells were initially used which contained large numbers of nMOS transistors with the p-channel devices being built directly in the substrate. This approach was the natural development of the existing CMOS technology. For the new circuit requirements, however, it was more appropriate to invert the polarities of the technology so that it becomes an extension of the standard n-channel process using a p-type substrate in which n-type wells are created for the pMOS transistors. Several companies have now adopted the inverted n-well approach. For example Intel have successfully produced parts such as a 4K static RAM and in the very large ROM category, Sharp has commercially available a 256K bit chip made in 3µm n-well CMOS which dissipates only 0.1µA in standby mode. The former, which uses a six transistor memory cell, is smaller than the equivalent HMOS II nMOS part and requires only a modest increase in process complexity.

Manufacturer	INTEL	HITACHI	TOSHIBA 1	TOSHIBA 2
Technology	nMOS	HiCMOS II	CMOS	CMOS
Chip Size mm^2	35.73	35.04	44.34	35.16
Gate Length µm	1.7	2	2	2
Active Power µW	400	200	15	15
Standby Power µW	100	–	5×10^{-5}	10^{-2}
Access Time nS	50	65	80	65

TABLE 3.5   Comparison of 64K static RAMs

In the field of random access memories, where process technology is at its most advanced, table 3.5 compares the performance and technology of some of the recent 64K static memories. All use two layers of polysilicon for interconnections, the CMOS versions using twin cell processes. The difference between Toshiba 1 and 2 is that in the former a more complex 6 transistor memory cell is used to minimise the standby power. The capability for low power is amply demonstrated by both Toshiba parts.

As with nMOS technology, scaling of CMOS devices is also desirable to reduce silicon area and increase speed. Predictably this affects the performance of the p and n-type devices differently and to optimise the circuit operation it is necessary to tailor the channels of each device independently. Hence the so-called "twin" well process is being used in which the doping levels in the n-type well and the p-type well are optimised independantly by implantation to reduce the second order effects and protect the devices against punch through and breakdown (Fig. 3.6b). The use of a lightly-doped epitaxial layer and an n+ substrate helps the twin well process overcome one of the drawbacks of earlier CMOS technologies - the tendency for devices to latch up through the formation of parasitic npnp and pnpn thyristors. This can be avoided by applying a small positive voltage to the substrate to short out the parasitic pnp transistors.

The success of the latest twin well CMOS technology has been demonstrated by Bell Laboratories with their Bellmac 32-bit microprocessor and by Toshiba in their 64K static RAMs. With an access time of 80 nS or less they have so low a power consumption that they will challenge the electronically erasable PROMs (EEPROM), particularly in applications which benefit from fast updating.

In spite of the Toshiba developments EEPROM technology is becoming widely adopted in place of the earlier ultraviolet-erasable EPROM technology for changeable non-volatile store. A 20-25V pulse is usually required to write data into an EEPROM or erase it while the normal 5V signal will read the memory contents. SEEQ Technology have however produced a 16K part which uses only a 5V signal for reading, writing and erasing. The device has all the high voltage generation circuitry integrated onto the chip.

The EEPROM technology uses a floating gate with tunnelling of charge through the insulating oxide from a conventional MOS control gate. This provides a changeable yet non-volatile memory under electronic control. To date most devices have been based on nMOS designs but CMOS is now being introduced. The new technology appears to be both commercially viable and more technically stable than the earlier electrically alterable (EAROM) technology which used a thin nitride insulating layer in the so-called MNOS technology. Most companies have ignored this technology although Fujitsu have recently announced a 2K EAROM using their Nitride Barrier Avalanche Injection (NAMIS) technology which runs against the trend. Another approach towards the ideal memory part which provides both non-volatility and changeability is the Xicor NOVORAM (non-volatile RAM). It combines a standard RAM with an EEPROM device which takes a "snap shot" of the RAM contents as required. The high voltage for the read cycle is generated on chip and the user need only provide normal 5V logic signals. Present parts are manufactured using a 5μm process which limits the capacity to 256 x 4 but larger memories are promised with technology scaling.

An inherent limitation in the speed of CMOS (and nMOS) is the capacitance incurred in the substrate and improvements can be made by using insulating substrates. The most popular to date has been sapphire and a number of companies notably RCA, Hewlett-Packard and GEC have developed silicon-on-sapphire (SOS) technology. In SOS the devices are constructed in a thin hetero-epitaxial layer of silicon grown on the sapphire. The devices are easily isolated by removing the silicon from between them and because parasitic capacitances are consequently much lower than in bulk devices, higher speeds are achieved. Using the same minimum feature sizes as in MOS, SOS devices are about twice as fast as bulk devices and are completely free from latch-up, even at the smallest geometries. Because of the simple method of isolation and the possibility of direct connections devices can be packed into 30% less area on the chip. For many of the second order effects which limit the amount by which scaling is possible, SOS devices are less sensitive and it presents an attractive option for advanced VLSI circuits. The major drawback, however, is the price of the substrate which is approximately an order of magnitude more expensive than bulk silicon. This is not too important in advanced custom chips as demonstrated by Hewlett Packard who have used the technology for some time. It is interesting, however, to note that Toshiba have experimented with SOS for their 16-bit microprocessors and 4K static RAMs which are aimed at the high volume merchant market.

A number of experimental SOS devices have been produced and RCA, who have pioneered the technology for many years, have produced an 8K CMOS/SOS non-volatile RAM with an overall access time of less than 40 nS. The main application for SOS and other Silicon-on-Insulator (SoI) technologies is in military systems where the radiation hardness of the technology is of importance.

3.5 FUTURE TRENDS

The main thrust of technological development is towards smaller geometries. At the research interface, laboratories are experimenting with devices drawn by electron beams directly on silicon with gate lengths down to 0.1 μm. Closer to the commercial world, effort is being applied to solving the CMOS latch-up problem as geometries approach 1 μm. Many of the major manufacturers have announced devices in development using such a technology and quote typical gate delays of 100-200 pS. As mentioned earlier the key to reducing the time delays of polysilicon interconnection and for the reduced gate lengths is the use of silicides and refractory metals. Tungsten and Titanium are the preferred materials and the techniques for depositing and patterning them are under development. In order to overcome topological problems with interconnection, multiple layers of metal and polysilicon are also being used with low temperature polymide insulating films.

Improved methods of device isolation are also under development. One approach is selective localised epitaxial growth on top of a bulk substrate which has been patterned with oxide. More advanced is the concept of building active devices on top of oxide, in recrystallised polysilicon. The manufacturing technology is still under development but several devices, including a basic RAM, have been produced. Interesting research at MIT Lincoln Labs has demonstrated a high-speed ring oscillator with inverter delays below 2ns built in 5μm CMOS on insulator technology. One interesting possibility of the recrystallisation approach is that it opens the door to three dimensional circuits and these are under investigation in a number of laboratories.

In the bipolar field improvements in speed is the main

driving force. Smaller device sizes with, particularly, shorter base widths are being produced. There do not seem to have been any fundamental breakthroughs recently and it is more a matter of gradual improvement along well-defined lines. The performance gap between bipolar and CMOS is narrowing rapidly and although the former is ahead in terms of pure speed at the present time, if the problems of scaling CMOS to below 0.5μm can be solved at an economic manufacturing price, that situation may not last much longer.

REFERENCES

1.   Moore, G.E., 1979, IEEE Spectrum 16, No 4.

2.   Millman, J., 1979, "Microelectronics", McGraw Hill.

3.   Garrett, L.S., Oct. 1970, IEEE Spectrum, 7, 44-56, Ibid. Nov. 1970, 63-72, Ibid. Dec. 1970, 41.

4.   Hart, K., and Slob, A., 1972, IEE J. Solid State Circuits, SC-7, No. 10, 346-351.

5.   Lohstroh, J., 1979, IEEE J. Solid State Circuits, SC-14, No. 3, 585-590.

6.   Murphy, B.T. et al, 1969, Proc IEEE, 57, 1523-1527.

7.   Forbes, B.E., 1977, April, "Hewlett Packard Journal", 2-8.

8.   Davidsohn, U.S., and Lee, F., Proc IEEE, 57, 1532-1537.

9.   Capece, R.P., 1979, Sept. 13, Electronics International, 109-115.

Chapter 4

Structure of semi-custom integrated circuits

P.J. Hicks

4.1   INTRODUCTION

Semi-custom integrated circuits can be classified under several headings according to their internal structure and the design style appropriate to their individual characteristics.   The purpose of this chapter will be to examine the structure and organisation of the devices which fall into each category and thus provide the necessary background for the material that follows.   As a first step it is clearly helpful if we can identify what the different types of semi-custom IC are.   In practice they can practically all be classified as belonging to one or other of four main groups. These are :

    (i)   Gate arrays (or Uncommitted Logic Arrays - ULAs)
    (ii)   Cell-based systems
   (iii)   Programmable logic devices (Matrix logic)
    (iv)   Analogue component arrays.

Of these the first three are almost exclusively digital with only group (iv) being specifically intended for analogue circuit applications.   However, it is possible to find semi-custom ICs that are capable of offering both analogue and digital circuitry on the same chip - a feature that can be particularly useful in applications requiring a single chip solution with an interface to the analogue world.   It may be that certain devices have characteristics belonging to more than one of the above categories - for instance some gate arrays are organised on the cell-based principle.

In the sections that follow each of the semi-custom design styles referred to above will be described in some detail.   Where possible, comparisons will be made in terms of criteria such as performance, packing density and suitability for particular applications.

4.2   GATE ARRAYS

The underlying concept on which all gate arrays or Uncommitted Logic Arrays are founded is of a chip containing a regular matrix of logic gates or components which is pre-processed up to the final stages of metal layer patterning. Approximately 90% of the wafer processing steps have therefore been completed before the chips reach the stage where

they will be committed to a particular function. The individual gates or components can be connected together in a
unique way to realise any required function, often with only
a single masking step. The principle is illustrated in
Fig.4.1. This shows the bringing together of preprocessed
silicon wafers carrying the uncommitted gate array chips and
a metallisation mask with the interconnection layout pattern
which will determine the eventual function of the chip
according to the logic developed by the designer.

Gate arrays can be broadly separated into two groups,
these being (a) medium to high speed gate arrays and (b)
very high speed gate arrays. The former are intended primarily for general-purpose logic replacement applications in
the industrial, telecommunications and consumer sectors of
the electronics market. They aim to cut production costs by
replacing, for example, an entire printed circuit board full
of SSI-MSI logic with a single semi-custom gate array
tailored to exactly meet a given product specification. A
wide variety of arrays is available in many different technologies including CMOS, NMOS, I^2L (with numerous variations),
CDI (collector-diffusion isolation) and Schottky TTL.

Very high speed arrays, on the other hand, are of more
interest to mainframe and large minicomputer manufacturers
since they achieve sub-nanosecond gate delays through the
use of technologies such as emitter-coupled logic (ECL).
The economics of producing this type of component are often
quite different from those in the first category. Mainframe
manufacturers, for example, are uncompromising in their
demands for the ultimate in terms of high speed performance
and are prepared to use very complex processes with potentially low yields. Other potential application areas are in
the field of high-speed digital signal processing.

4.2.1 Technology

Complementary MOS (CMOS) technology is proving to be
very popular with gate array manufacturers. Metal-gate MOS
technology was used almost exclusively in these arrays until
recently, although they are now being superceded by faster
and denser silicon-gate CMOS structures. In particular,
oxide-isolated silicon-gate CMOS processes with their superior packing density and switching performance are now being
adopted by several of the gate array vendors. Many of the
commercial CMOS arrays are currently fabricated with 5 μm
minimum feature sizes, although the next generation of
devices made with 3 μm design rules is now emerging. At
these dimensions gate delays of the order of 3 ns are attain-
able, speeds that were once unimaginable for a CMOS gate.
Furthermore, whereas single metallisation layers have
normally been used in the past, the more modern arrays are
moving towards the use of double-layer metal to achieve
higher on-chip packing densities.

The disadvantage of double-layer metallisation is the
more complicated processing required in order to customise
the array. Instead of using just a single mask to pattern
the one metal layer, typically two extra masks are needed

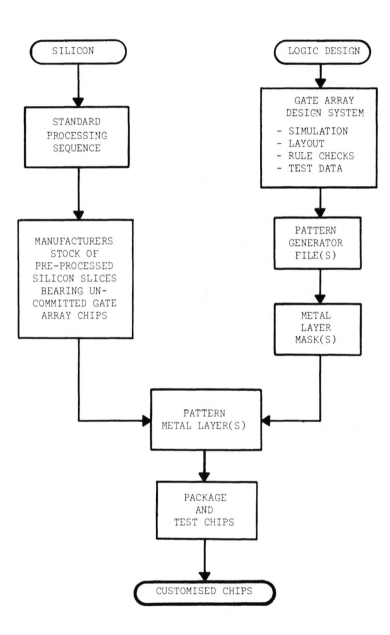

Fig. 4.1   The gate array principle

for a two-layer metal process. One of these provides the
interconnection pattern for the second layer of metal, while
the other is used to produce contact holes or 'vias' through
the insulating layer that separates the two levels of metal.
Offset against this extra processing complexity is the fact
that automated layout of the chip becomes considerably
easier to achieve if two levels of interconnect are available rather than one.

Silicon-gate CMOS arrays containing around 10,000
gates are already being produced and even bigger chips are
planned. As device geometries are scaled down to $1\,\mu$m and
beyond it will become technically feasible to make VLSI gate
arrays with 100,000 gates or more. When this point is
reached, however, the problems will no longer be with the
technology, but in finding a methodology that will enable
complex designs to be successfully integrated onto such
large chips.

Of the bipolar technologies used for manufacturing
gate arrays, Integrated Injection Logic (I^2L) offers a very
low power-delay product and high packing densities. Gate
delays are of the same order as for CMOS and dynamic power
consumption at high switching speeds can actually be lower
than for that technology. There are a number of I^2L variants, for instance ISL (Integrated Schottky Logic) which
combines the packing density of I^2L with the high speed of
Schottky TTL. Bipolar CDI technology has been used by one
array manufacturer (Ferranti) to produce RTL and CML
(Current Mode Logic) ULAs for a number of years. The gates
can be optimised for either high speed or low power operation and recent improvements in the technology have made it
possible to produce ULAs containing 10,000 gates.

ECL arrays are offered by several of the big semiconductor IC manufacturers (e.g. Fairchild, Motorola and
Plessey), and others are produced in-house by computer manufacturers and are primarily intended for internal use. A
typical example might contain over 400 gates and use two or
even three layers of metallisation for orthogonal tracking
and power distribution. The power dissipation of the larger
ECL arrays can reach many watts and presents a severe packaging problem.

An important advantage that several of the bipolar
array producers have been able to offer is the ability to
integrate both analogue and digital functions onto the same
chip. Some CMOS arrays are now also able to provide this
capability.

4.2.2 Organisation

Considerable variations are possible in the overall
organisation or architecture of gate arrays as well as in
the detailed structure of the cells themselves. Three basic
array organisations can be identified, although some detail
differences are bound to exist even within these categories.
Simply stated the divisions are :

    (1)   The Block-Cell approach - the cells are spaced
           apart at regular intervals on a two-dimensional
           matrix (see Fig.4.2a.). The vertical and

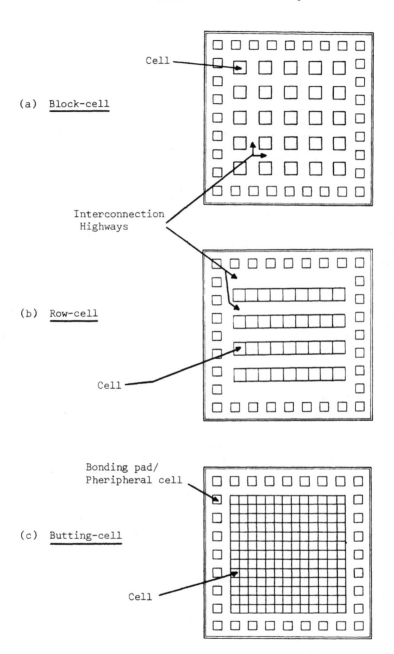

(a) Block-cell

(b) Row-cell

(c) Butting-cell

Fig. 4.2   Organisation of gate arrays

horizontal channels created by the gaps
between the cells are available for routing
the interconnections between cells.

(2)   The Row Cell approach - in this case the
cells are set out in horizontal or vertical
rows with fairly wide gaps between the rows,
as shown in Fig.4.2b.   These gaps are again
provided for intercell routing.

(3)   The Butting Cell approach - here the cells
abut one another and no gaps are left for
routing.   Instead, sufficient room is available for the routing of interconnections
through the cells themselves by making
allowance for this in their design.   The
basic principle is illustrated in Fig.4.2c.

In every case the regions around the edge of the chip are
dedicated to peripheral cells that are responsible for
interfacing the internal logic to the outside world.   In
many cases these input and output functions will involve
level conversion; for example, between TTL logic levels (0
and +5V) and whatever logic levels the circuitry inside the
array uses.   Tri-state and open-collector type outputs are
also usually available, and in some cases functions such as
Schmitt triggers, oscillators, monostables, etc., may be
included.

Generally speaking categories (1) and (2) above permit
automated wiring between the cells to be carried out using
two levels of interconnection for orthogonal tracking (i.e.
tracks on the two different levels run at right angles to
one another).   These levels may consist of a single layer of
metal and another of either diffusion or polysilicon in the
form of fixed underpasses or crossunders.   Alternatively the
two levels may be two separate layers of metallisation with
contact holes or vias to allow connections to be made between them.   Examples of these two wiring schemes are shown
in Fig.4.3.   By contrast the wiring of arrays in the third
category is almost invariably done manually at the moment,
the reason being the added complication of having to route
the wires through the cells themselves.   To facilitate the
crossing of tracks it is again usual to provide crossunders
in either diffusion or polysilicon at fixed locations within
the cell structure.   As one might expect the packing densities in terms of logic per unit area are normally higher for
arrays of the third type than for the other two.

The detailed structure of the cells is a function of
the technology in which the array is fabricated as well as
the organisational style it uses.   In most cases the cells
consist of a small group of transistors and other components
arranged in such a way that they can be easily interconnected to form functional logic elements.   Some of the components may already have common connection points as a built-
in feature of the cell : for example, a pair of bipolar
transistors may share a common collector, or MOS transistors
may have common sources and drains.   Apart from these built-
in connections the components are otherwise 'uncommitted' in
that their role in the final circuit is not determined until
they are interconnected by tracks in the metal layer.

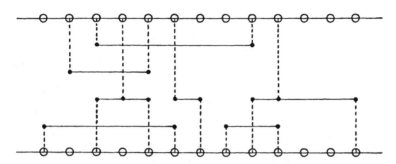

(a) This diagram shows how the routing of tracks can be achieved using two metal layers. The lower layer (dotted) runs only in the vertical direction while the upper layer (solid) is constrained to the horizontal.

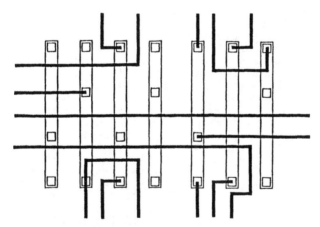

(b) Crossunders in polysilicon or diffusion placed at fixed, pre-arranged locations on the chip are used here to achieve pseudo two-layer routing of interconnections (metal tracks are drawn as solid black lines: the squares represent contact holes through to the crossunders).

Fig. 4.3   Orthogonal routing of interconnections on two layers

A few examples of different cell designs will be given to illustrate various possible internal arrangements. The diagram in Fig.4.4a., for instance, shows a typical layout for a CMOS cell belonging to the row-cell category. The contents of the basic cell comprise two p-channel and two n-channel MOSFETs with extended polysilicon gates which also serve as crossunders through the cell. In addition a large number of contact holes are provided at preset locations to enable access to be gained from the metal layer to the diffusion (source and drain) and polysilicon levels. Finally, a couple of metal power lines pass through the cell to distribute $+V_{DD}$ and ground across the array. In order to link these components together to form a simple logic element such as a two-input NAND gate it is necessary to add a few internal connections to the cell on the metal layer. These are often referred to as intracell connections and one way in which the components can be wired to form a CMOS NAND gate is shown in Fig.4.4b. The tracks between cells which link the various, gates together are known as intercell connections and are routed in the channels provided for this purpose (see Fig.4.2b.). In arrays where only a single layer of metallisation is available the intercell metal track will run horizontally along the channels and terminate on contacts to fixed polysilicon crossunders which run vertically across.

In contrast to the CMOS cell described above, the layout shown in Fig.4.5a. belongs to an Uncommitted Logic Array fabricated in bipolar CDI (Collector Diffusion Isolation) technology. Each of the array cells contains a group of uncommitted transistors and resistors that can be connected together to produce typically a pair of two-input NOR gates. The individual components within the cell in Fig.4.5a. can be related to the circuit schematic of Fig.4.5b which represents just one of the pair of CML gates that can be made. A typical layout of the interconnecting tracks needed to produce a pair of the gates in Fig.4.5b. is shown in Fig.4.5c. Note that the cells in this array abut one another with no wide channels left for routing. The organisation of this array therefore fits into the third category of the classification scheme defined earlier. To make it possible to connect up the cell components and then wire between them each cell contains three diffused crossunders (labelled XU in Fig.4.5a.). Both of the power supplies for the cells in this array are distributed through the bulk silicon, thus avoiding the need to sacrifice precious routing space on the single metal layer for this purpose. Every cell has one access point to the positive supply rail V_S and one to the ground rail GND.

It is clearly not possible, in the space available, to describe all the different gate array cells that exist. Arrays fabricated in ECL technology, for example, normally have cells that are quite a lot larger than the two discussed so far. A unit cell in an ECL gate array may contain as many as 20 or 30 uncommitted components (transistors, resistors and diodes) which when connected together will produce a logic function such as a two-input multiplexer.

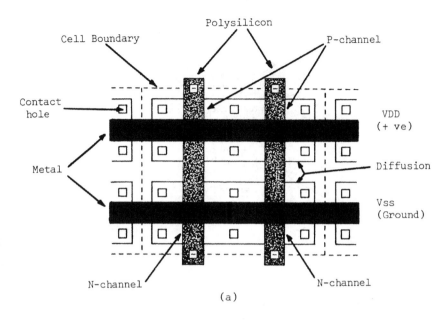

(a)

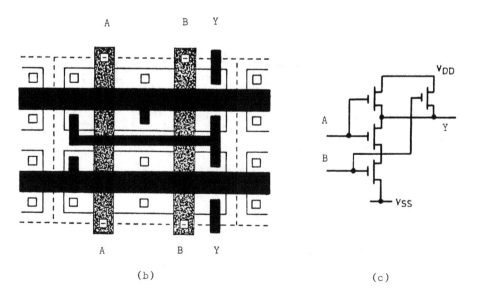

(b)

(c)

Fig. 4.4 (a) Layout of a CMOS gate array cell of the row-cell type. (b) Internal connections required to produce a 2-input NAND gate, (c).

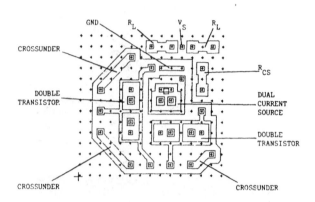

(a) Matrix Cell

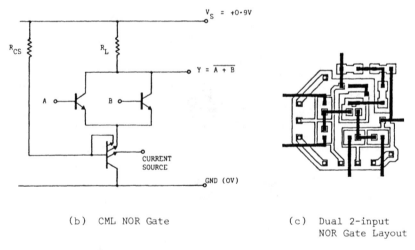

(b) CML NOR Gate

(c) Dual 2-input
      NOR Gate Layout

Fig.4.5   A single matrix cell from the 440-cell ULA described in the
text is shown in (a). The circuit schematic of a 2-input CML NOR gate
appears in (b), and the metal interconnections required to produce a
pair of such gates from one of the matrix cells can be seen in (c).

(Courtesy of Ferranti Electronics Ltd.)

Cells in an I^2L gate array will be different again, reflecting to a large degree the physical layout characteristics peculiar to that technology.

An interesting idea for a gate array based on the principle of the Universal Logic Gate (ULG) has been proposed by Hurst (1). Universal Logic Gates are combinational circuits which can be made to perform all possible Boolean functions of a given number of input variables. (Yau and Oric (2)). The function generated depends on the permutation of the connections made to the inputs of the ULG. For example, a ULG capable of realising all functions of two input variables would be termed a ULG.2 circuit and has three input terminals and one output. Internally the ULG.2 circuit will comprise three or four simple gates connected in a two level AND-OR configuration. The advantages claimed for ULGs as functional gate array components stem from their complete universality. Whereas it is often necessary to generate an all-NAND or all-NOR equivalent of a logic circuit before it can be mapped onto a conventional gate array, the ULG array overcomes the need for this. It does suffer from the disadvantage, however, of having redundancy built into every cell.

4.3   CELL-BASED SYSTEMS

This technique requires the production of a full mask set, typically needing as many masks as a full custom design. It exploits the hierarchy inherent in all logic systems by partitioning the logic into functional building blocks or cells. The cells themselves are custom-designed and held in a cell library. For a given design appropriate cells are selected from the library and placed in rows across the chip. Sufficient space is left between the rows to allow for interconnections between the cells. An example of a typical standard cell layout is shown in Fig.4.6. Cell-based systems represent the only really automated CAD layout style which is fully and successfully used in custom LSI production at present. The organisation of a standard cell IC with its regularly-spaced rows of cells and broad routing channels has been deliberately chosen to facilitate layout using existing placement and routing algorithms. In fact the algorithms used are identical to those that have traditionally been applied to the automated layout of printed circuit boards. As a general rule it is necessary to have two levels of interconnection available for such routers to operate effectively, although they can be made to work with one.

4.3.1   Technology

The fabrication technology that seems to have been almost universally adopted by the manufacturers of standard cell ICs is CMOS. All the comments that have been made previously concerning this technology apply equally here.

The reasons why CMOS is preferred over other technologies for cell-based custom ICs can be traced back to its

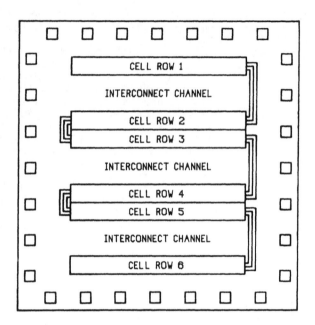

Fig.4.6.   Organisation of a Standard Cell IC.

reputation for low power consumption and a near-ideal gate
transfer characteristic, the latter being responsible for
the high noise immunity and excellent fan-in and fan-out
properties which make CMOS so easy to work with from a
designers point of view.

4.3.2 Organisation

As can be seen by comparing Fig.4.6. with Fig.4.2,
standard cell ICs share the same overall chip organisation
as the row-cell approach to gate arrays.
A standard cell library might contain of the order of
fifty different logical functions chosen from an existing
logic family - for example, the CMOS 4000 series SSI/MSI
range of ICs. Once the layouts for the cells have been
fixed on all mask levels, they can be readily called up agair
and again from the library to suit the requirements of any
design to be integrated. The procedure is analogous to the
use of macros in an assembly language programme; each time
the macro is called, the group of machine instructions that
it represents are inserted into the programme. In most
cases the cells are all constrained to be a fixed height but
can vary in width according to the complexity of the function that each represents.
Since the cells themselves are custom-designed it is
not possible to comment further on their internal structure.
The only thing that can be added is that the cell connection
points emerge from the top and bottom edges which interface

with the routing channels.

The automatic placement and routing packages for standard cell ICs are capable of laying out chips containing several hundred cells with 100% efficiency and consume relatively small amounts of CPU time in the process (of the order of minutes). The price that must be paid for this fast turnaround is a very inefficient use of silicon area - the density of a circuit implemented in this way may only be half that of a custom chip designed to perform the same function. Attempts to improve the utilisation of silicon area, for example by making the routing channels narrower, immediately lower the routing efficiency to possibly 60-70% or even less. The connections that the router was unable to insert then have to be added manually - a lengthy and tedious business. The algorithms used for placement and routing will be considered in detail in Chapter 12.

It is worth investigating how current cell-based architectures perform as chip complexity approaches VLSI. De Man (3) has shown that as the number of cells N becomes greater the CPU time required increases as N^{1+K}, $0 \leqslant K \leqslant 1$, for placement algorithms and approximately linearly for fast channel routers. However, the most significant problems arising from increased complexity are to be found elsewhere. They are :

(1) As row length increases the routing space grows quadratically with N if the cells are randomly interconnected. This rate of growth reduces to linear in N if the placement of cells could be ordered in such a way that only local interconnections had to be made, although in reality the best that can be expected is probably $N^{1.5}$ dependence.

(2) There is a tendency for wires to bunch around the middle of the routing channel resulting in congestion. This problem is aggravated as the channel length increases and leads to further loss in density.

(3) Standard cells are fairly well-suited to the implementation of random logic functions but far less able to cope satisfactorily with bus-oriented systems or regular structures such as ROM, RAM, PLA, etc. Both of these features are commonly found in VLSI chips.

The conclusions to be drawn from the above observations are that standard cells can certainly provide a highly-automated and therefore rapid means of producing custom chips, assuming that a fairly high proportion of the total silicon area can be given over to routing the cell interconnections. Using existing algorithms the fraction of area lost to routing will get larger as chip complexity increases, and if the technique is to survive as a VLSI methodology it will be necessary to build more degrees of freedom into the system. One way in which this could be done would be to invoke hierarchical placement and routing whereby groups of pre-designed cells can be assembled and interconnected to form larger cells of arbitrary size. These can be combined yet

again to form still larger cells, and so on. Since the
blocks can be of arbitrary size there would appear to be no
objection in principle to including regular structures such
as ROM, RAM and PLA if desired. In fact custom LSI chips
combining ROM and RAM arrays with the conventional standard
cell style of architecture have already been produced,
notable examples being from Bell Laboratories.

The fundamental problem associated with cell-based
systems is that the intercell wiring is divorced from the
densely-wired cells themselves. This results in a dispro-
portionately large fraction of the total chip area having to
be devoted purely to wiring and can be directly attributed
to the fact that the wiring is effectively added as an afterthought. The only real way of overcoming this is to embed
the logic in the wiring, a strategy which at present is best
illustrated by highly regular structures such as ROM and PLA.
To accomplish this successfully it is necessary to plan how
the chip will be wired in advance, at the so-called 'floor
planning' stage. One of the cornerstones of the structured
design methodology for VLSI proposed by Mead and Conway (4)
is founded on this principle.

4.4   PROGRAMMABLE LOGIC DEVICES

This heading is applied to a family of components, the
members of which are characterised by a logical function and
interconnection topology inherent in their array-like structures. They are all basically capable of implementing arbitrarily complex logic functions in sum-of-products form, and
vary mainly in the degree of programmability afforded to the
designer. Sequential logic can be realised in the form of
finite state machines with the addition of appropriate
external feedback connections. The devices are programmed
to suit a particular application by either leaving intact or
removing physical connections (such as fusible links) in the
matrix.

The best-known examples in this category are the
familiar PROM (Programmable Read Only Memory) and FPLA (Field
Programmable Logic Array). More recently the family has
been extended with the introduction of Programmable Array
Logic (PAL), the Field Programmable Gate Array (FPGA) and
Field Programmable Logic Sequencer (FLPS). Since all these
devices are described in greater detail in later chapters
only a brief outline need be provided here.

The overall architecture of the PROM, FPLA and PAL is
shown schematically in Fig.4.7. Note the similarity of the
logic function performed by each array, the differences
being determined by whether the AND matrix, the OR matrix or
both the AND and the OR matrices are programmable.

From the designers point of view the task of programming these components is a straightforward extension of
conventional logic design and requires no special IC design
or layout skills. The design cycle is short and the cost
therefore fairly low, even for one-off quantities. However,
the rigidity of the logic function tends to limit the range
of applications, and in many cases there is a fair amount of
redundancy to take into consideration. Another feature that

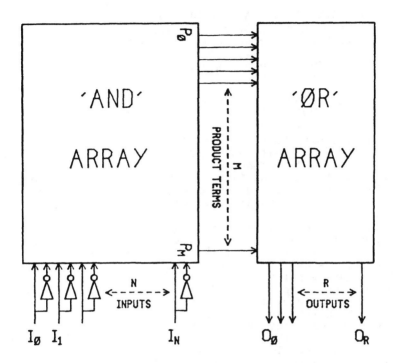

Fig.4.7 A comparison of programmable logic devices.

(a) PROM : 'AND' array pre-programmed. Fully decodes N inputs
to yield all M = 2^N possible product terms.
'OR' array user-programmable. Any combination of
the M products can be ORed onto any of the R outputs.

(b) FPLA : 'AND' array user-programmable. Generates a chosen
subset of M ($<2^N$) product terms from N inputs.
'OR' array user-programmable. Any combination of
the M products can be ORed onto any of the R outputs.

(c) PAL : 'AND' array user-programmable. Generates a chosen
subset of M ($<2^N$) product terms from N inputs.
'OR' array pre-programmed. Groups of product terms
are ORed onto the R outputs according to a pre-arranged
pattern.

some users might consider undesirable is the ease with which
the contents of a programmed logic component such as a ROM
can be copied. This means that a design implemented in this
way will tend to have a low security rating.
A new variant based on the matrix logic concept and
known as the Storage/Logic Array (SLA) has been described in
the literature by Goates et al (5). The SLA is essentially

a PLA with interleaved AND and OR planes but has the added
facility of permitting storage (memory) elements to be
embedded within the array where required. It possesses many
of the attributes required of a structured design methodology
for VLSI and in most cases can provide higher circuit densities than the PLA alone.

4.5   ANALOGUE COMPONENT ARRAYS

These are intended for purely analogue functions and
invariably employ bipolar technology, although recent developments in CMOS design have resulted in linear arrays being
introduced in this technology also. A typical analogue
array uses standard junction-isolated bipolar technology and
contains from around 100 to 500 components. These comprise
both pnp and npn transistors of various sizes, and a selection of resistors.

Analogue building-blocks such as operational ampli-
fiers, oscillators and comparators can be readily assembled
by suitably interconnecting the components. The designer
usually begins by breadboarding the circuit he wants to
integrate using parts supplied by the array manufacturer.
These have characteristics identical to those in the array
itself. Having proven the design the next stage is to
manually translate this into a layout on a suitable array.
This is done on a layout sheet (also supplied by the manufacturer) showing the location of all the components superimposed on a special wiring grid at 200 x full size.
Bearing in mind that a single metal layer has to be patterned in order to provide the component interconnections,
pencil lines are drawn on the layout sheet where these paths
are required.

By taping over the pencil lines a 200 x master artwork
for the metallisation mask is produced and this is then
photographically reduced to generate a 10 x reticle. A step
and repeat camera provides the working masks for the metal
etching step which dedicates the preprocessed array slices
to the desired function.

4.6   SUMMARY

The aim of this chapter has been to review the various
different types of semi-custom IC that are available. As
such it sets the scene for the later material which deals
with the steps involved and the tools used in designing with
these devices.

The layout styles of gate arrays and standard cell ICs
means that they are currently well-suited to 'mopping-up'
chunks of random logic that might otherwise require 20 to 30
standard off-the-shelf SSI/MSI ICs. They do not perform
well, however, in situations demanding substantial amounts
of on-chip memory (e.g. RAM, ROM) or regularly-structured
logic such as PLAs. Their ability to cope adequately with
bus-oriented logic is also somewhat limited. Since these
are all characteristic features of VLSI chips it is reasonable to assume that further evolution of the structure and

organisation of semi-custom ICs will occur in future.

REFERENCES

1.   Hurst, S.L., 1980, 'Custom LSI design : the universal-
      logic-module approach', Proc. IEEE Int. Conf. on
      Circuits and Computers, 1116-1119.

2.   Yau, S.S. and Orsic, M., 1969, 'Universal Logic
      Modules', Proc. 3rd Princeton Conf. Information Sci.
      Sys., 499-502.

3.   de Man, H., 1981, 'Computer-Aided Design Techniques
      for VLSI', in 'Design Methodologies for VLSI Circuits'
      ed. Jespers, Sequin and Van de Wiele, NATO ASI series
      no. E47, Sijthoff and Nordhoff, 1981, 113-174.

4.   Mead, C. and Conway, L., 1980, 'Introduction to VLSI
      systems', Addison-Wesley.

5.   Goates, G.B., Harris, J.R., Oettel, R.E. and
      Waldron, H.M., 1982, 'Storage/Logic Array design :
      reducing theory to practice', VLSI Design, III, No.4,
      56-62.

Chapter 5

Selection of semi-custom technique, supplier and design route

J.R. Grierson

5.1   INTRODUCTION

Whenever a manufacturer develops a new piece of electronic equipment, be it a space rocket or a child's toy, he should at some point consider whether some or all of the electronics should be integrated onto one or more silicon IC's. His first choice is thus whether to leave the circuit as developed (probably in off-the-shelf SSI/ MSI), or have it integrated using a custom or semi-custom technique. On what basis he should make this choice is considered in Section 5.3

If he decides on one of the three main semi-custom techniques he must next decide who will supply him with the IC and Section 5.4 lists some of the questions that must be asked in the making of that choice. Finally having chosen a technique and a supplier the customer must decide to what level he wishes to become involved in the design and the choice of entry point to the design system is considered in Section 5.5.

All three choices are based on the customers own priorities and needs, thus there are no clear-cut answers. The intention of this chapter is to highlight some of the advantages and disadvantages that may be encountered.

5.2   SEMI-CUSTOM TECHNIQUES

For the purposes of this chapter, semi-custom techniques will be divided into three broad classifications, Programmable Devices, Gate Arrays and Cell-Based Systems. These are briefly defined below:-

(a) Programmable Devices
This covers all techniques where no technology processing is required to customize development samples, and will include FPLAs (with and without registers), PALs (with and without registers) and PROMs (UV or electrically programmable). With all these techniques mask programmable options are available for production. They have no linear capabilities.

(b) Gate Arrays
This covers all techniques where only a few masks
(typically 2-4) are required for device customizing. The basic uncommitted components are always
pre-processed onto the slice prior to customizing.
Both digital and analogue arrays are available as well
as mixed analogue/digital.

(c) Cell-Based Systems
This includes all systems where a full mask set is
required for all customizations. It thus covers a
range of techniques including fixed height cell,
variable height cell, poly-cell, symbolic logic,
gate matrix etc. This classification has probably
the widest variation with various combinations used,
and in the limit becomes full custom design.

5.3   CHOICE OF SEMI-CUSTOM TECHNIQUE

In order to make a choice of which semi-custom
technique to use for a particular application numerous
factors must be taken into account. The priority of these
will depend on the application but they may usually be
broadly grouped as under:-

(a) Total cost (development cost, unit cost, cost of
associated re-design etc).
(b) Timescales (time from initial design to
production of chips).
(c) Design technique capability and flexibility.
(d) Ease and accuracy of design to first time success.
(e) Technology capability (can the technology meet the
speed, power consumption etc requirements).

Item (e) is a vital issue, but generally separate from the
others in that within each semi-custom technique it is
possible to use a range of technologies. If the ultimate
in speed or power consumption is required full custom
design must be used, however for more normal applications a
suitable pairing of design technique and technology can
usually be found.
The choice of semi-custom technique is thus mainly
concerned with the groupings (a) to (d) above and these
will be considered in relation to semi-custom techniques,
SSI/MSI and full custom.

5.3.1 SSI/MSI Technique. This is the classic 'breadboard'
technique, often used for the development of the circuit.

(a) Cost - The development cost may be very low
indeed if the initial system design was breadboarded.
If automated assembly is to be used some further
development will, however, be needed. In production the unit cost remains high, almost independent
of the volume (Fig 5.1)

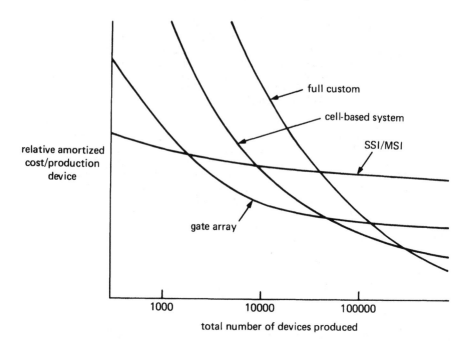

The positions of the abscissa scale points are the subject of constant debate. Those shown above are averaged from the literature.

Fig 5.1 The cost-volume curves for various design techniques.

(b) Timescales - As with cost, the timescale to first prototype may be very small. However the time to produce each board, even with automated assembly will be long and independent of volume.

(c) Flexibility - The technique is extremely flexible and most functions can be incorporated.

(d) Ease and accuracy - Design is usually straightforward but checking the final product for accuracy and conformance with the design is not easy. However modifications are readily made.

One of the overriding disadvantages of SSI/MSI is of course the large amount of space it consumes and in addition it will normally be a power-hungry solution.

5.3.2 Programmable Devices . Although one of the oldest semi-custom techniques programmable devices are confined to a rather small area of application because of their relative inflexibility. However their cost and timescales make them attractive for some applications.

(a) Cost - Development cost is very low, a few man
days of effort to construct the program and then the
use of a programming machine which itself costs only
a few thousand pounds. In production the chips are
only a few pounds each, though the unit cost should
be compared on a cost/logic function basis.

(b) Timescales - Development timescales are a few
days and often only hours to correct a fault. In
production a mask-programmed device can be produced
in large volume.

(c) Flexibility - Only a very limited range of
functions is available.

(d) Ease and accuracy - Development is normally a
straightforward job and first time success is common,
though this is partly due to the relatively limited
capability. Faults are easily corrected.

Overall programmable devices are quick and cheap to develop
and produce but have a very limited application range.

5.3.3 Gate Arrays. By far the largest sector of the
semi-custom market is that of the gate array. They are
currently available in virtually all modern technologies
and in sizes up to 8000 gates.

(a) Cost - the development cost of a gate array is
the sum of the design costs and the prototype fabrication costs. The design costs will depend on the size
of the circuit and array, the circuit complexity and
the design aids available. With the latest design
aids the design cost should be kept to a few thousand pounds provided a reasonably well-proven circuit
is being integrated.

The main selling point of gate arrays is the low
prototype fabrication costs. With the uncommitted
silicon produced in large volume, only the metal
masks to be made for each circuit and only the metal
to pattern, prototype fabrication costs can be kept
much lower than a technique requiring a full mask
set.

The unit cost in production of a gate array depends
on the chip costs and the package and test costs.
As gate arrays are of fixed sizes not all the gates
will normally be needed, hence a gate array chip
size will be larger (and therefore more expensive)
for the same technology than the cell-based or full
custom chip of the same circuit.

The overall cost however must include the amortized development cost and thus the true cost of any I.C. is given by:-

$$Cost = \frac{D}{N} + Chip + F$$

...... (5.1)

where D = Development cost
        N = Number of chips manufactured
        Chip = Unit chip cost in production
        F = Packaging, testing etc costs per chip.

The gate array has a relatively low 'D' and high 'Chip' compared with a full-mask set system, thus tends to be economic at smaller volumes (Fig 5.1).

(b) Timescales - the length of the design phase of a gate array development will depend entirely on the starting point. With a good, well proven design, translation to gate array logic and simulation could be done in a week or two, and automatic layout could be completed in days. However, test program generation and verification will still take several weeks. On the other hand if the design has hazards or is badly thought out, it could take months to modify it into a form suitable for a gate array. Mask making and fabrication could be completed in two or three weeks but most manufacturers quote about 8 weeks to allow for queuing, scheduling and checking of the mask-making, metal patterning, packaging and testing.

As a rough rule-of-thumb, for an 'average' array from a 'reasonably well-proven' circuit (though not in gate array logic) and allowing for the various customer/supplier interfaces, about six months should be allowed as a realistic development time.

(c) Flexibility - gate arrays can accommodate virtually any purely digital function on chip, with the exception of memory (RAM or ROM) components which are so wasteful of gate array area as to be totally impractical, unless specifically dedicated on the chip. Pure analogue arrays are available which can be formed into a wide range of analogue functions. Some mixed analogue/digital arrays are also available but in general these are fairly limited in scope, can be wasteful in area and are rarely well supported in terms of CAD, simulation etc.

(d) Ease and accuracy - the availability of computer aids for gate array design has increased dramatically over the last few years and fully automated systems are becoming available. A customer should now expect a supplier to have a comprehensive range of computer aids available working to a common database,

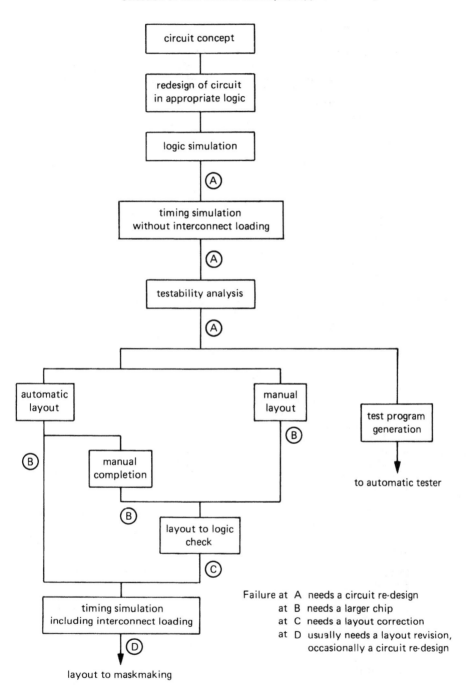

Fig 5.2 Design path for a gate array or cell-based system

and   these must include:-

(i) Simulation - at least a full timing analysis
with worst case parameters, and fan-out and layout
dependent delays. Higher level simulators are starting to be introduced.
(ii) Testability analysis - usable before the design
is finalized.
(iii) Layout - either guaranteed fully automatic or
with full checking if manual intervention permitted.
(iv) Test program generation and verification -
automatic generation is unusual and often inefficient
but verification is essential.

Overall gate arrays are widely available in a large
range of options, are heavily supported, reasonably
flexible and are relatively cheap and quickly
available. Thus they are very useful for prototyping
and the lower production volumes, though it should
be noted that some gate arrays are in very high
volume production.

5.3.4 Cell-Based Systems. The main rival to the gate array
in the field of 'silicon-customized' systems is the cell-
based system. As mentioned in Section 5.2 there are a
wide variety of such systems, though most of their characteristics can be considered here to be similar.

(a) Cost - As with a gate array, the use of
automated layout means that the design portion of the
development cost will be dominated by the circuit/
simulation costs and will be similar to a gate array
design cost. The fabrication costs to prototypes
will be considerably larger than the gate array costs
as a full mask set must be produced and processed.
At about £1000 per mask, mask-making costs alone can
increase prototype costs considerably, in addition to
the higher overheads of small batch runs.

In terms of silicon area, the layout of a cell-based
design is more efficient than a gate array in the
same technology as each chip is only the minimum size
required for the circuit and there are no unused gates.

Using equation 5.1, the development cost D is higher
for a cell-based system than a gate array, but the
cost per working die is less (more dice per wafer). F
will be roughly constant for the same circuit and thus
the amortized cost will be lower than gate arrays at
higher volumes (N). (Fig 5.1).

(b) Timescales - Development timescales follow much
the same comparison with gate arrays as do the costs.
Design times will be similar, though for non-auto-
matic layout cell-based design times were often
shorter. As all masks must be made and all stages

processed, fabrication times to prototypes are longer than for gate arrays. As always however fabrication times will tend to be dominated by queues and
scheduling.

(c) Flexibility - Cell-based systems can in theory be
completely flexible by using pre-designed cells for
effectively any function, digital, analogue, memory
etc. In practice the problems of maintaining the
cell library limits the library to what is manageable.
The cell-based system is thus more flexible than the
gate array but in practice cannot have the total
flexibility of full-custom.

(d) Ease and accuracy - CAD was generally available
for cell-based system before it became available for
gate arrays, particularly in automated layout. Flexibility may be constrained by the need for full CAD
(eg mixed analogue/digital simulation is not common),
but the same programs are needed as for gate arrays,
ie those listed in Section 5.3.3.

Generally cell-based systems are more suited to higher
volume applications than gate arrays. However the increasing sophistication of both types of system is leading to
fierce competition between the systems over a wide range of
production volumes.

<u>5.3.5 Full Custom Design</u>. The classic full custom design
technique, where every transistor is designed, optimized
and laid out by hand, is used for a decreasing proportion
of circuits. However, as the other extreme of silicon
integration it is useful to compare its attributes with
semi-custom.

(a) Cost and Timescale - The overriding feature of
full custom is the long and hence costly manual
design phase. Several man years of effort are
normally expended on this for a fully hand-crafted
design. Fabrication costs will be for a full mask
set as in cell-based design, but these costs are
usually dwarfed by the design cost.

The prime aim of full custom design is usually to
achieve minimum chip size (As well as fully hand-
crafting this can sometimes be by 'improving' a cell-
based design). In terms of equation 5.1, 'D' is high
but 'chip' is low, thus amortized cost is the lowest
of all systems at very high production volumes
(Fig 5.1). Full custom is often only economic for
'standard parts'.

(b) Flexibility - Another reason for the use of full
custom is its total flexibility. It is the only type
of silicon integration which is genuinely totally
flexible, though clearly it can also be mixed with a

cell based system for a faster, flexible system.
In some cases full custom may be the only solution
for a particularly difficult problem (eg mixed
technologies on a chip).

(c) Ease and accuracy. Full custom is a difficult
and highly specialized activity. Because of the
total flexibility, CAD tools tend to be large and
complex, but even then they are usually not as
'fool-proof' or comprehensive as semi-custom tools,
leading to a higher probability of error.

Because of timescales and development costs hand-
crafted full custom design is giving way to semi-
custom design. Only where chip size is of extreme
importance or where mixed design modes are needed is
full custom likely to be used for future design, and
even these parts of the chip may be designed by
semi-custom techniques.

TABLE 5.1 League table for semi-custom techniques

Ordering (1 best, 5 worst) is done on the basis of implementing a total logic system in silicon using only one
semi-custom technique, plus standard parts where the semi-
custom technique cannot cope.

	SSI/ MSI	Programmable Devices	Gate Arrays	Cell Based	Full Custom
Development cost	1	2	3	4	5
Unit cost in production	5	4	3	2	1
Size of logic system	5	4	3	2	1
Power consumption	5	4	3	2	1
Number of devices	5	4	3	2	1
Development time	1	2	3	4	5
Flexibility for design	1	5	4	1	1
Modification cost and time-scale	1	2	3	4	5

5.4   CHOICE OF SEMI-CUSTOM SUPPLIER

The choice of supplier will depend on the technique chosen, the entry route selected and the technology required. In some cases, eg PLA's there is a fairly limited range of suppliers and the choice may be made in a fairly straightforward manner. However in other cases, such as gate arrays there is a large and increasing number of suppliers.

The choice of technology is of course vital to any selection, however assuming that choice has been made, a list of suppliers able to supply devices in that technology can be drawn up. The following questions (in arbitrary order) may be useful for a customer to use in choosing a supplier from that list.

(i)   What design routes are available? This question is dealt with in more detail in the next section, but an appropriate one must be available.

(ii)   What computer aids are available for design? If for supplier use, what checks has the customer that they have been used? If for customer use, how easy are they to use, what access does he have to them? A minimum set of computer aids is given in Section 5.3.3.

(iii)   How much engineering support does the customer have if doing his own design?

(iv)   What design documentation is available for customer-design? Are all design parameters clearly stated, worst case, backed by practical results and included in all design tools?

(v)   How clear is the customer to supplier interface? What is the customer contracted to provide for the supplier? Is the form, quantity and quality of this information fully and unambiguously described? Does the supplier provide a written statement of acceptance?

(vi)   How clear is the supplier to customer interface? What is the supplier contracted to provide for the customer? Usually the first supply will be a specified number of packaged devices. Are these devices tested - functionally?, parametrically?, to what test program?, under what conditions?

(vii)   Who owns the intellectual property rights on the design? What are the conditions on retention, ownership and release of masks, guarantees of supply?

(viii) For gate arrays, who will supply the uncommitted silicon, the customizing masks and the metallization processing? For cell-based systems who will supply the processing?

(ix)   Who will provide the packaging and testing? Are there different suppliers of the processes for prototype and production chips? Are there second sources for any or all of the processes?

(x)   How many "satisfied customers" has the
       supplier? (This can be a very sensitive
       question due to the confidential nature of
       customer/supplier relations. Suppliers should
       never release what work has been done, but they
       may be allowed to release names of whom they
       have worked for. The number of designs completed
       may not be a useful guide if a large proportion
       of such designs are for the supplier's in-house
       customers).

(xi)   How much will the development cost? What will
       be the unit cost of the devices in production?
       Is there a minimum order quantity associated
       with acceptance of the development contract?

(xii) What are the timescales to prototype chips?, to
       production chips? How firm are those time-
       scales, can they be proved to have been
       achieved in the past? How much do they depend
       on outside suppliers?

(xiii) What packaging options are available?

(xiv) What Q.A. (Quality Assurance) procedures are in
       place? To what Q.A. specifications are the
       final devices produced? What reliability
       information is available for these types of
       devices?

It is not suggested that this list of questions is complete
and certainly the priority attached to the questions will
depend on the application. However the main problem is
often interpreting the answers. In particular, aspects
such as ease of use of computer aids and adequacy of design
documentation tend to be subjective assessments and until
they have actually been tested by the customer it is very
difficult to give a realistic judgement. In this area
experience is very valuable and may be an influence on the
entry point chosen.

5.5   CHOICE OF ENTRY POINT AND DESIGN ROUTE

The choice of customer's entry point will depend on
the semi-custom technique used and the customer's individual circumstances. While programmable devices offer a
relatively small choice, the more complex gate array and
cell based designs have a wider choice.

5.5.1 Programmable Devices.
PLA's ROM's etc can be either
electrically or mask programmable. For prototype work
electrically programmable versions are normally used. If a
customer foresees only a small number of designs, programming of the devices is readily available from the supplier.
However many customers find that the relatively low cost of
programming machines, and the very fast turnround possible
with your own machine makes customer-programming an
economic route for prototype devices, even with relatively
small numbers of designs.

5.5.2 Gate Arrays and Cell Based Designs. For these
systems there are a variety of interaction points between
customer and supplier. At one extreme the customer does all
the work up to fabrication, at the other he hands over the
design to the supplier at the initial concept stage. In
addition to these two a third party can be used - the
independent design house.

5.5.2.1 The independent design house. The design house can
act on a customer's behalf through some or all of the design
phases, some points when considering the use of a design
house are:-

  (i)   The design house may act for the customer when
         the customer does not have the appropriate
         resources and when the supplier is unwilling to
         commit his design resources to a design that is
         not guaranteed to go into production.
  (ii)   The design house should be able to assist with
         the choice of supplier. However the customer
         must exercise care here as many design houses
         are closely tied to one or two suppliers and
         their advice may not be totally impartial.
  (iii) The design house is paid only by the customer
         and if required will spend more time optimizing
         a design. A supplier's prime aim is to get a
         device into production and he may not be willing
         to supply extra resources (even when paid) on
         detailed optimization.
  (iv)   A design house will be willing to take an overall
         system-level view of a design and decide the best
         means of system implementation.
  (v)   Generally a design house can operate very
         effectively on behalf of an inexperienced
         customer. However the customer must ascertain
         the design house's experience in the chosen semi-
         custom technique and its access to in-house or
         supplier's CAD tools.

In the rest of this section reference will be made only to
the customer and supplier. It is assumed that in any stage
the customer could be replaced by a design house acting on
his behalf.

5.5.2.2 The interface points. Fig 5.3 shows eight possible
interfaces between customer and supplier applicable in gate
array or cell system development. Moving from interface 1
to 7 involves the customer in increasing own-manpower
commitment and decreasing payment to the supplier. This is
generally favoured by suppliers, involving them in less
speculative work and less commitment of design resources.

Interface 1. This interface, involving the supplier in
possibly speculative system design, raises problems of
specification and definitions of boundaries. It is not a
good interface for customer or supplier.

Interface 2. This is the classic 'breadboard' interface when a working, debugged board of SSI/MSI is to be reproduced in LSI. This is usually a 'clean' interface and a precise I/O specification can be written. Due to technology differences, however, timing problems can arise unless a careful timing specification is given. The high level simulator interface can eliminate the latter problem, but has the disadvantage that unlike the bread-board it cannot be tested as a direct replacement in a working system.

Interface 3. Redesign of the circuit in terms of the available logic functions can help to optimize the design (eg use of NAND rather than NOR gates in CMOS) and eliminate timing problems if in-house simulation is available. The interface will consist of a logic diagram (graphical or textual) with timing information and I/O specifications. This is a preferred interface for many suppliers where the CAD tools are only available to the supplier. It is the last stage at which significant logic changes can be made.

Interface 4. Logic simulation, timing simulation, test-ability analysis must all be applied to the circuit. If the customer uses his in-house simulators he will need the timing data from the supplier and this may not be easily incorporated. If he uses the supplier's simulators he must learn how to use them. If the supplier takes over at this stage it is probable the customer will still have to check the results and he would certainly have to make any circuit modifications found necessary. It is likely that both customer and supplier will inevitably be involved in this stage.

Interface 5A. Test program generation (TPG) is currently one of the least automated, and consequently most difficult parts of semi-custom design. Some customer involvement is invariably required to provide at least some initial test vectors, and many suppliers insist on the customer supplying the full test program. Automatic TPG is rarely effective in general circuits and the cost of supplier-generated test programs should encourage customers to consider test-ability all through the design process.

Interface 5B. In contrast to TPG, a great deal of automa-tion is available for layout. If a supplier provides fully automatic layout, a customer would be well advised to use it. On the other hand manual layout is largely learnt from experience. If a customer intends to manually lay out several designs he may wish to learn the art, if not he is probably advised to leave it to the experienced supplier. If any manual layout is used the supplier would be expected to provide automatic layout-to-logic checking.

Interface 6. If manual layout was used outside a graphics system, the layout must be digitized for mask making.

Unless the customer has numerous designs with one supplier to justify loading the design database onto his own graphics system, this stage is almost certainly best left to the supplier. With automatic layout, or layout directly on a graphics system the data base is normally incorporated and no digitizing is needed.

Interface 7. Only in rare cases is a supplier likely to accept masks from a customer. Problems of mask compatibility, alignment marks, process control chips etc make mask level interfaces very difficult.

Occasionally the supplier may provide untested devices to the customer if the customer has his own ATE though this can lead to difficulties if the customer claims to find zero yield - who is at fault?

Undoubtedly the appropriate interface will depend on the customer and supplier. In addition to the information given above, other technical data such as packaging, pin-out, voltage range, temperature range etc will always have to be provided. Most customers will probably interface at points 2, 3, 4 or 5 and will usually move from 2 towards 5 as their experience increases. Whichever interface is chosen, however, the customer must ensure that there is a clear statement of responsibilities in technical and financial terms if problems arise later in the development. Potentially this can be one of the biggest areas of difficulty in semi-custom developments.

5.6 SUMMARY

The choice of semi-custom technique, supplier and design route will depend on the customer and his particular requirements. Every situation is different and this chapter seeks only to give guidance and to point out some of the strengths and weaknesses of various approaches. Three main criteria must always be satisfied:-

(a) The final device must meet all technical requirements (functionality, reliability etc).
(b) The total cost of obtaining the devices must make the end product economically viable.
(c) The timescales to obtain the devices must be acceptable in respect of the production of the end product.

ACKNOWLEDGEMENT

Acknowledgement is made to the Director of Research of British Telecom for permission to make use of the information contained in this chapter.

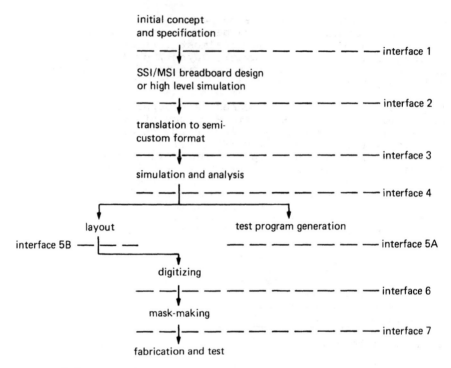

Fig 5.3 Possible customer-supplier interfaces for a gate array or cell-based design.

Chapter 6

Circuit design techniques

J.W. Tomkins

6.1 INTRODUCTION

This chapter examines Gate Array and Standard Cell semi-custom design techniques and their interaction with circuit design at the transistor level. Circuit techniques available to the full custom designer are presented and the constraints which are applied in the production of semi-custom designs are discussed.

6.2 CUSTOM DESIGN

A custom design converts a functional and timing specification into a realisation in silicon. In general the design will progress with three major aims in view :-
   (i)   to achieve the desired speed and functionality,
   (ii)   to achieve a design having minimum production costs,
   (iii) to achieve the function for the minimum power consumption.
These three requirements in combination will dictate the choice of both process and circuit techniques. The requirement for minimum production cost further implies that the final design should be as compact as possible.
   Having chosen the process, the full custom designer is free to use any combination of component type and value that the process can offer. In practice however the area efficiency varies greatly between the various components available on a given process and this biases component choice towards a subset of the range available.
   Figures 6.1 and 6.2 show the relative sizes of the components available on bipolar and MOS processes respectively. As can be seen there is a distinct advantage in the bipolar process towards the use of transistors and low value resistors. In the case of MOS processes the area advantage of the transistors becomes overwhelming, which accounts for the almost total absence of resistors in MOS circuit designs.
   The following two sections will review various circuit design techniques and examine their advantages and disadvantages in terms of area, speed, and power dissipation.

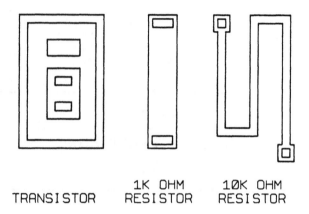

Fig.6.1   Relative sizes of a selection of bipolar
          IC components.

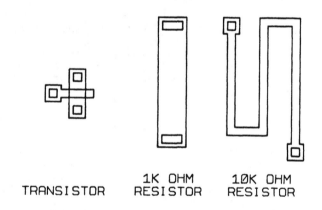

Fig.6.2   Relative sizes of a selection of MOS IC
          components.

6.3   DESIGN TECHNIQUES - BIPOLAR

Two broad categories of design technique will be
reviewed in this section, Resistor Transistor Logic (RTL)
and Emitter Coupled Logic (ECL). At first appearance this

may seem a peculiar combination, each however has been chosen
to illustrate a particular aspect of bipolar design. ECL
represents the high speed end of the spectrum, and also the
driving force behind the development of many modern bipolar
processes. RTL, on the other hand, although not a widely
used technique for custom design, is interesting in that it
has been used as the basis for some gate array families. In
this application it is instructive to see how the constraints
applied to produce an array have actually nullified some of
the disadvantages of the technique as a custom design tool.

6.3.1   Resistor Transistor Logic (RTL)

Resistor transistor logic is a very early form which in
terms of standard logic families has now been totally superceded. A basic two input RTL NOR gate is shown in Fig.6.3.

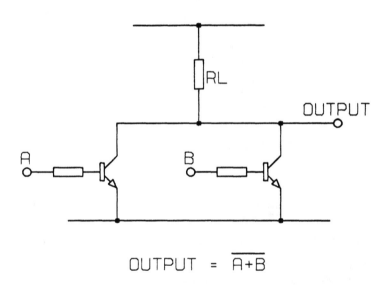

$$OUTPUT = \overline{A+B}$$

Fig.6.3   The basic 2-input RTL NOR gate.

With both transistors in the 'off' condition the output is
pulled high by the load resistor R_L. Should either of the
inputs be at a high level, however, that transistor will
pull the output to within a few hundred millivolts of ground.
In this condition the load resistor sees the full supply rail
applied across it. To minimise power consumption therefore
the resistor must have a relatively high value, probably in
the region of 10 kilohms or greater. To preserve fanout and
noise immunity the input resistors must be of a similar or
higher value. The problems with this structure are its low
speed, due to saturating transistors and high value resistors,
and the area penalty of using many high value resistors.

These problems can be alleviated by increasing the logic power of the gate by the use of more complex stacked structures as shown in Fig.6.4. In this example the logic power has been more than doubled whilst component count has risen by an appreciably smaller percentage. An additional advantage is that the power consumption has not changed, there still being a single load device, hence resulting in a substantial improvement in power-delay product.

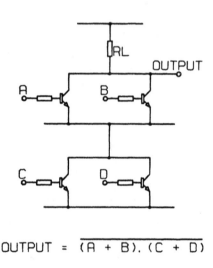

$$\text{OUTPUT} = \overline{(A + B).(C + D)}$$

Fig.6.4   An example of a stacked RTL gate, in this case producing an OR/OR NAND function.

Further improvements in packing density and power consumption can be achieved by the addition of multi-emitter transistors used as emitter followers on the output of the gate. This arrangement increases the noise immunity of the gate and, more importantly, allows the construction of the wired-OR gate. This is achieved by simply wiring together the emitters from several gates hence providing a multi-input OR gate at very little cost in area and for no increase in power consumption.

Even with the incorporation of all these improvements, however, the gate is still relatively demanding in area terms due to the use of the resistors. This disadvantage, however, can virtually disappear in circuits using single layer metal or in gate array applications. When interconnecting gates to form a full system there is always a need for interconnect crossovers and these are provided free in the case of RTL as each input has its own resistor which can be used as a cross-over. In the case of gate arrays it is usual to leave routing highways, free of contacts, between rows of gates. With this sort of structure the resistors can be snaked under the otherwise empty routing highways either passively, consuming otherwise under-utilised area, or actively where they can be used to provide interconnect crossovers.

6.3.2 Emitter Coupled Logic (ECL)

Emitter coupled logic currently provides the fastest logic form available in standard silicon technology. Simple ECL, although achieving very low gate delays, suffers two disadvantages - component count is very high, as can be seen from the two input NOR gate of Fig.6.5, and the gate is rela- tively power hungry. Both of these limitations are overcome to some extent by circuit variations which increase the logic capability of the gate significantly whilst having smaller impacts on device counts and power consumption, as with the RTL examples. Two commonly used techniques are the wired-OR and the wired-AND.

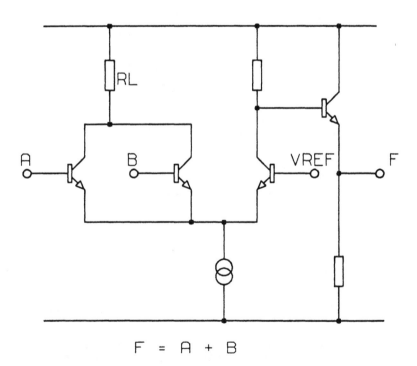

$$F = A + B$$

Fig.6.5 A simple ECL OR/NOR gate.

The wired-OR is the easiest to construct, involving only the connection of two or more outputs to a single input.

This arrangement realises a multi-input OR function whilst
requiring no extra components. A single output can be used
in a number of wired-OR's simply by adding extra emitters to
the output emitter follower, one per wired-OR.
    The wired-AND function is achieved by shorting together
internal nodes on a number of gates as shown in Fig.6.6.
For the addition of a single additional clamp diode this
arrangement realises a multi-input AND gate which again
considerably increases the logic power of the gate. Using
unmodified ECL NOR gates the example of Fig.6.6 would have
required an extra four transistors, a resistor, and a
further current source with an attendant 50% in power consumption.

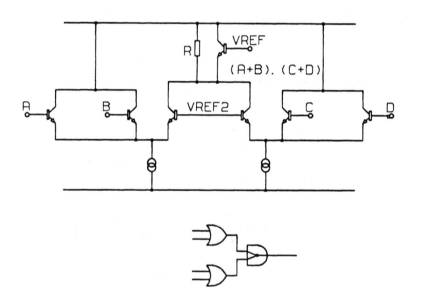

Fig.6.6   The ECL wired-AND function

    The logic capability of the basic gate can also be
increased dramatically by 'stacking' gates as shown in
Fig.6.7. The design and tolerancing of this type of circuit
is more complex as the gate now requires two reference voltages, separated by about one volt, and the lower tier logic
inputs also require level shifting by the same amount. In a
design operating from a 5 volt supply it is possible to use
three tiers of gating resulting in a large reduction in the
number of current sources with attendent reductions in power
consumption. The example of Fig.6.8 shows the complexity of
a master-slave flip-flop realised using multi-tiered ECL.
The stacked design requires approximately one-fifth the

power of the non-stacked version.

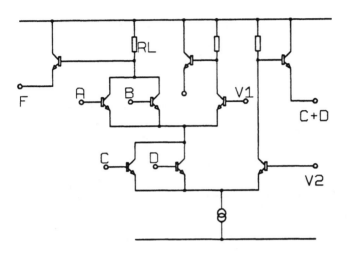

$$F = \overline{(A+B).(C+D)}$$

Fig.6.7   'Stacking' allows complex functions to be
generated economically in ECL

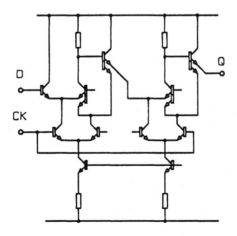

Fig.6.8   An ECL master-slave flip-flop

In practice combinations of these power and device saving techniques will be used in a full custom design. The circuit of Fig.6.9 shows a parity tree which uses all of the techniques to produce the function :

F = ABCD + A'B'C'D' + ABC'D' + A'B'CD + AB'CD' +
    AB'C'D + A'BC'D + A'BCD'

at the expense of only two current sources.

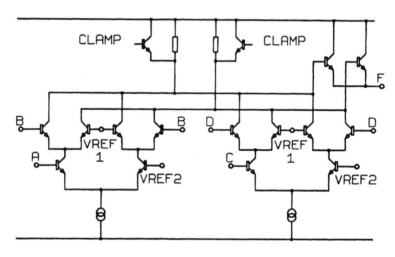

F = (ABCD + A̅B̅C̅D̅ + ABC̅D̅ + A̅B̅CD + A̅BCD̅ + AB̅C̅D + A̅B̅C̅D + A̅B̅CD̅)

Fig.6.9   A parity tree implemented in the form of an
          ECL complex gate

For larger scale integration a reduction in component count is essential and can be achieved at the expense of on-chip noise margins. Standard ECL has a logic swing of around 800mV which provides sufficient noise margins for connections between chips on a printed circuit board and even between boards in a system. Within the more controlled confines of the chip itself, however, the swing can be safely reduced to 450mV.

This small change is enough to have a dramatic effect on the circuitry. The reference voltages in a stacked structure need only be separated by one base-emitter voltage drop, V_{be}, and the output emitter followers, which previously provided level shifting, may be dispensed with. In addition it is possible to introduce an extra upper level of logic which is useful despite the limitation that only the inverse output of this level is available due to saturation problems. Further improvements in packing density can be achieved by introducing multi-emitter transistors. The effectiveness of

these measures can be judged from the clocked D type latch
of Fig.6.10 which requires only a single current source,
five transistors, and a resistor. Additionally the input to

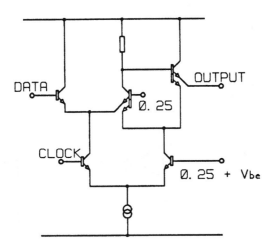

Fig.6.10   An efficient realisation of a clocked
D type latch in ECL

output delay is equivalent to one gate delay, which for a
modern 3 micron process is of the order of 0·5nS at 1mA tail
current.

6.4   DESIGN TECHNIQUES - MOS

MOS processes, whilst lacking the capability to provide
useful resistors, do offer other features which can be
exploited to advantage in digital integrated circuit design.
The availability of depletion mode transistors in NMOS processes, and complementary devices in CMOS, allow the use of
compact transistors to replace the bulky resistor loads used
in bipolar circuitry, whilst the extremely high input impedance of the devices allows direct interconnection of gates
with a total absence of current hogging problems. In addition the high input impedance allows the use of dynamic
techniques which can give large improvements in power consumption.
The following three sections will examine static,
dynamic and CMOS design techniques and will discuss some of
their relative merits.

6.4.1   MOS Static Logic

The structure of a basic NMOS static NOR gate is shown
in Fig.6.11. As can be seen the circuit is analogous to the

value '1' or '0'. If S subsequently switches from true to
false, the pass transistor in series with the D input will
be turned off and the Q output will instead be connected to
the input of the first inverter via the pass transistor
controlled by S̄. In this condition the logic state at Q
will be continuously regenerated by virtue of the positive
feedback connection.

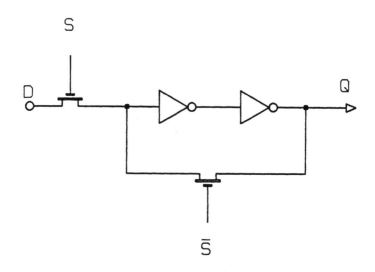

Fig.6.12 Using pass transistors to form a simple
data latch in NMOS technology

Although these techniques do reduce the overall power
consumption there is still a considerable dc content which
in general dominates over the dynamic power required solely
to move the circuit nodes at the required rate. This dc
content can be removed by the use of dynamic design techniques.

6.4.2 MOS Dynamic Logic

Dynamic techniques exploit the high 'off' resistance of
the MOS transistor coupled with its isolated gate to produce
a circuit form having virtually no dc power consumption.
This is essentially achieved by replacing the permanently
conducting depletion load device of static NMOS with a
clocked load. To totally remove any dc current consumption
the input transistor, which perform the logic evaluation,
are clocked in anti-phase to the load (or pre-charge) device,
hence eliminating the dc path between the supplies. It
should be noted that static NMOS (and its bipolar counter-
part, RTL) relies for its successful operation on the estab-
lishment of a suitable ratio between the resistance of the

RTL gate, the resistive load being replaced by a depletion mode NMOS transistor connected so that it is permanently conducting and the input resistors being omitted as there is no current hogging problem to contend with. The total elimination of resistors has overcome the area problem that occurs in RTL but the logical power of the gate is still weak and the structure still consumes dc power.

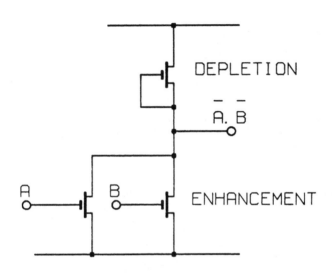

Fig.6.11   The basic NMOS static NOR gate

As might be expected from the similarity of the circuit structures, similar modifications can be used to improve the logic functionality/power consumption trade off. All of the stacking techniques discussed in relation to RTL logic can be applied, with similar effect, to the MOS static structure. In addition to these techniques, however, the symmetry of the MOS transistor coupled with the total isolation of the gate electrode from the channel makes possible a further useful circuit innovation, namely the pass transistor. In this application the MOS transistor is used to connect the output of one gate to the input of the next under the control of an enabling signal. In this way it becomes easy to produce compact multiplexers requiring the use of a single MOS device per input line. An example of the use of pass transistors to form a multiplex function is shown in Fig.6.12 where a two transistor multiplexer has been used in conjunction with two inverters to form a D latch. When control signal S is true the data input D is connected to the input of the first inverter. The gate capacitance at this node will be charged to almost the supply voltage V_{DD} if D is at logic '1' or will be discharged otherwise. The output Q of the second inverter correspondingly takes on the

pull-up load device and the 'on' resistance of the pull-down
transistor switch. This ratio is important for the definition of a satisfactory logic low level at the gate output,
and for this reason logic circuits of this type are often
grouped under the generic heading of 'ratioed' logic.
Dynamic logic circuits, on the other hand, require no such
attention to resistance ratios and hence are commonly
referred to as 'ratioless'.

Fig.6.13 shows a complex gate converted to dynamic form
by the addition of two clocked transistors. The two clocks
CK1 and CK2 are non-overlapping and in antiphase. When CK1
is high the output node is connected to the positive supply
and hence charges to within one threshold voltage V_{TH} of
that supply. The clock polarities are now changed and the
output node is conditionally discharged depending on the
logic states applied to the evaluation transistors. Provided
the inputs do not change the output node will remain valid
until the following clock period when the output will be
pre-charged ready for the next evaluation cycle.

To prevent the inputs changing whilst the gate is evaluating a further pair of non-overlapping clocks are provided
and alternate logic states are timed with alternate clock
pairs as shown in Fig.6.14, which shows one stage of a
dynamic shift register. As can be seen from this example a
further benefit arises in that each output node provides a
free storage medium, data being stored in the form of charge
trapped on the nodal capacitance. It is this storage mechanism which accounts for the simplicity of the shift register
example of Fig.6.14.

The main advantages of dynamic logic are its low power
consumption, which automatically adjusts to the clock speed,
and the ease of providing large amounts of storage at very
little cost. Disadvantages are the number of clock phases
to be generated and distributed and the fact that alternate
stages must run from alternate clock pairs hence increasing
wiring complexity and the chance of mistakes during layout.
An additional restriction is caused by the dynamic storage
inherent to the technique. Leakage from circuit nodes will
eventually destroy the logic value stored on a node and this
effect sets a lower limit on clock frequency if correct
operation is to be guaranteed. The limit is usually in the
tens of kilohertz region.

6.4.3 Complementary MOS (CMOS)

By providing complementary transistors the CMOS process
allows the use of a design style which combines the insensi-
tivity to clock frequency of the static structure with the
low power consumption of the dynamic approach. This is
achieved by replacing the depletion load of the NMOS static
gate with complementary P channel transistors as shown in
the examples of Fig.6.15. The P channel transistors are
connected to perform a logic function which is the inverse
of that performed by the N-channel transistors. This
arrangement guarantees that there is always a dc path
between the output and one of the supply rails but never a

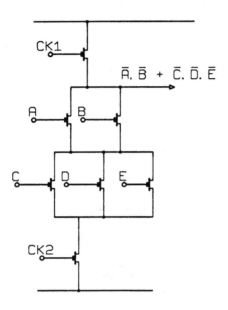

Fig.6.13   An example of an NMOS complex gate using dynamic logic

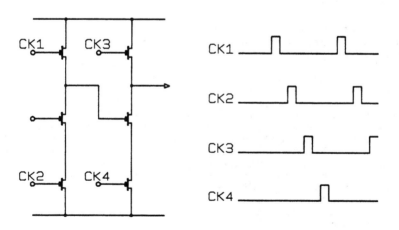

Fig.6.14   A single stage of a 4-phase dynamic shift register

path directly from one rail to the other. CMOS thus provides a further example of 'ratioless' logic.

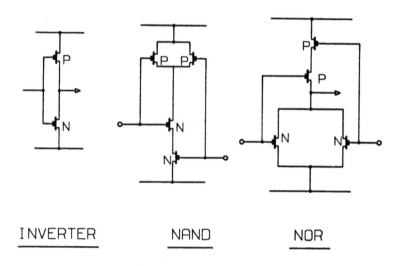

INVERTER   NAND   NOR

Fig.6.15 A selection of simple CMOS logic functions

The complementary gate structure provides static logic structures which exhibit purely dynamic power consumption. The price paid for this desirable combination, however, is increased component count over the NMOS depletion load equivalent. Additionally this increase in components per logic function is not improved significantly by the use of complex gate structures as is the case with NMOS structures. This is because every additional device in the N evaluation tree has to be complemented by a further P channel device.

The use of truly complementary structures has its worst impact on large structured logic blocks such as PLA or ROM. The situation is made worse in these array structures due to the increased clearances required between N and adjacent P transistors to accommodate the isolating well. To give improved packing density it is desirable that the array should consist of only one device type. This can be achieved, whilst avoiding dc power consumption, by the use of a dynamic array structure supported by static peripheral circuitry. This combination vastly reduces component count whilst maintaining dynamic power consumption and low frequency clock operation.

A useful component that is available in CMOS technology is the transmission gate. This is effectively the CMOS equivalent of the NMOS pass transistor. It comprises two transistors, one P channel and the other N channel, connected in parallel with their gates driven by complementary control signals. Although more complicated than the simple

NMOS pass transistor, the transmission gate possesses
superior switching characteristics and can also be successfully employed as an analogue switch.

The application of dynamic techniques to CMOS designs
can be extended to random gate structures using the arrangement known as domino logic. A domino logic gate is shown in
Fig.6.16. In this structure the complementary load devices
are replaced by a single clocked load transistor and a
further clocked device is added in series with the evaluation tree, much as with the four phase dynamic logic of the
previous section. In the CMOS implementation however the
load is a P channel device and the series gate is an N type
device, thus allowing the use of a single clock for both
devices. The complex gate is followed by a static inverter,
thus producing an overall structure that is non-inverting.

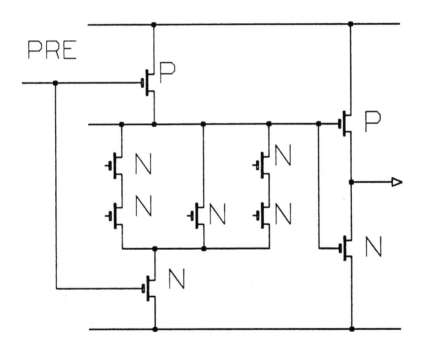

Fig.6.16 'Domino' logic utilises a pre-charging
          technique

The use of a non-inverting gate allows a number of
gates, all using the same clock phase, to be cascaded. In
the pre-charge state all evaluation nodes will be pulled
high and hence all outputs will be at a low level. At the
beginning of the evaluation mode the clocked loads go off
and the evaluation trees are enabled. All gates except the

first in the cascade, however, have lows on their inputs hence maintaining their outputs at a low level. Evaluation of the input logic state therefore ripples down through the cascade of gates without the need for intermediate clocking. This technique allows a reduction in device count whilst maintaining high speed operation and requiring relatively simple clock generation and distribution.

6.5 DESIGN TIMESCALES

The starting point of any design is a specification of the desired function. The chip designer expands and refines this definition into a more detailed description, partitioning the circuitry into lower level blocks.

During the partitioning designs will be favoured which give rise to repeated circuitry, regular structures such as RAM, ROM, or PLA, or previously encountered functions. Quite early in the partitioning phase the most suitable technology will be identified, high speed requirements favouring bipolar and high complexity/low power favouring MOS.

Once the technology has been decided the partitioning continues. Now, however, the designer will be considering the emerging design in terms of circuit level considerations such as ease of implementation of the functions in the technology, and area and power dissipation constraints.

The design will eventually be broken into logic level blocks and the design verified by simulation at this level. In a full custom design each logic block will then be further decomposed to produce a design at transistor level and each block will be simulated and optimised at this level. Following satisfactory simulation layout will begin, the components of each logic function requiring individual attention. Layout will be interleaved with re-simulation to check on the effects of stray capacitance introduced during layout. Finally the logic level blocks will be interconnected and the final layout checked before processing can begin.

The advantages of full custom are that the design of each block of circuitry is optimised for the application, both in terms of layout and in the choice of circuit design technique. In fact on a full custom chip it is possible to mix techniques such as ECL and I^2L to separately optimise performance of different parts of the circuitry. The disadvantage is the time taken to accomplish the circuit level design, simulation, and layout.

As a design is refined down from the specification the number of entities the designer has to manipulate grows. This growth is most pronounced during the conversion from logic to component level and it is this level that most semicustom techniques seek to eliminate in their quest for reduced design time and increased design security.

The two most common semi-custom approaches, gate arrays and cell-based methods, both seek to eliminate component level design from the chip timescales but do so by applying slightly different compromises to the circuit level problem.

6.6 SEMI-CUSTOM GATE ARRAYS

The gate array approach provides an array of predefined components laid out in a predefined pattern, usually consisting of block of components separated by empty routing highways. Design is supported by the provision of a library of predefined logic functions which can be implemented on the array. For each function the library will contain a description of the logic function, including timing, and a set of interconnect patterns which, when superimposed on a block of components in the array, will produce the desired function.

The system designer selects and specifies the interconnection of the predefined logic cells necessary to achieve his desired function. Having designed the function in terms of the library cells, layout becomes merely a matter of placing the cells on the predefined array of components and producing, either automatically or by hand, an interconnect pattern to connect the elements in the desired way.

This approach does, however, place severe constraints on the designer of the cells as he now has a limited number of components with a limited range of values from which to construct the given function. The situation is made even more restrictive in that any new cell should have interfaces (both in terms of timings and levels) that are compatible with all other cells in the library. To make the task possible at all a great deal of planning must go into the choice of components available, and into their relative positions and numbers within the array.

In general it will only be possible to optimise the choice of components to one design technique, which will then be used exclusively in all the cell designs on that array. The following two sections will examine the options available in ECL arrays and CMOS arrays.

6.6.1 ECL Arrays

Many ECL gate arrays are now readily available on the commercial market, each using slightly different circuit techniques internally. In general, however, the CML type of construction with three levels of current switching appears to be the favourite in the 300-3000 gate range. The reasons for the popularity of this technique should be clear from a reading of section 6.3.2. The stacked approach maximises the ratio of logic power to both device count and to power consumption whilst maintaining, and in some cases enhancing, the speed capability of ECL. A modern high-speed array will operate with tail currents in the range 0.5mA to 2mA and will achieve gate delays in the 0.5nS region.

The basic building block of an ECL array consists of 20-40 components which have been chosen and laid out to allow easy implementation of commonly used functions, such as flip-flops, multiplexers, and adders. The blocks are placed next to a predefined bias generator cell, as shown in Fig.6.17, and this unit is then repeated across the array in rows with routing channels between. The periphery of the chip is surrounded with bonding pads and special blocks whose components have been optimised for external interface use.

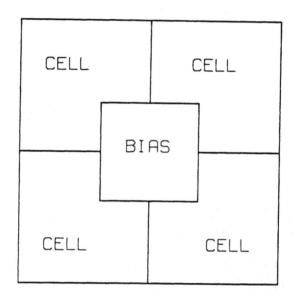

Fig.6.17   The overall arrangement of a cell cluster
in a typical ECL gate array

6.6.2   CMOS Arrays

The choice of circuitry for use in CMOS arrays is constrained by the requirement for ease of use and interfacing
of the functional cells seen by the systems designer.
Dynamic logic, whilst offering area advantages, needs
special care during design, clocks have to be correctly
phased, the capacitance of dynamic nodes can be critical to
correct operation, and there is a minimum clock frequency
below which the circuit will fail. These conditions would
be very difficult to control on a general purpose array, and
standard static logic is used in most arrays, the transistors
being configured to also allow ease of construction of transmission gates.
     As already stated in Chapter 4, nearly all currently
available CMOS arrays are constructed in a similar way. The
basic approach is again demonstrated in Fig.6.18. This cell
consists of two N transistors, two P transistors, and a
crossunder, the transistors being connected in pairs with a
common node. The arrangement of the contacts to the N and P
transistors will depend on whether the array is intended to
support the use of transmission gates. Where this is the
case the gate contacts will be separate, otherwise a common
contact to N and P gates can be used with consequent area

saving.

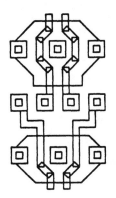

Fig.6.18   An example of a layout for a CMOS gate
array cell

The increasing complexity capability of modern MOS
processes is leading to modified array structures. To try
to improve the area efficiency of the array there is a move
towards the use of larger building blocks chosen to optimise
more complex cells such as RAM or MSI logic functions. This
approach can give improved efficiency but the problem of
producing array structures that are optimum over a range of
cell types increases as the complexity and diversity of
those cells increases.

6.7   SEMI-CUSTOM - CELL BASED

Cell based semi-custom systems differ from gate arrays
in that there is no predefined component group or chip
structure which must be used for all cells and all applications. The cell based system instead provides a library of
functional blocks each of which has been optimally designed
using whatever components were applicable to that particular
function. At the system design level, however, the approach
looks similar to that of gate arrays. The chip is designed
in terms of the standard functions available in the library
and its performance verified by simulation on a logic simulator. When the design is complete, the required cells are
placed onto the chip and are interconnected either manually
or automatically, depending on the system being used.
Freedom from the predefined array of components allows
a wider choice of design techniques to be used in the design
of the individual cells in the library. The only constraint
is that the cells conform to a standard interface specification, and have connection points that lie on a standard grid.
Many currently available cell based systems offer simple
cell libraries, and in these cases the design technique used

is consistent from cell to cell, with a static approach
usually being adopted. As cell complexities grow, however,
the design techniques will be chosen on a cell to cell basis
and it can be expected that the range of techniques used
will expand to cover virtually all those currently used in
full custom design.

Chapter 7

Logic design with emphasis on ASM method

E.L. Dagless

7.1   INTRODUCTION

Semi-custom and custom integrated circuits today provide, at best, a set of logic primitives which can be mani-
pulated to produce a desired design. However, the sheer
complexity of the potential circuit makes entirely intuitive, unstructured methods of design somewhat daunting.
Other parts of this course will address some existing and
future solutions to these problems. The purpose of this
chapter and a later one on PLA and ROM-based design is to
outline one method of logic design, the Algorithm State
Machine (ASM) method, that permits a degree of abstraction
away from the circuit description. The result is a design
method permitting considerable creativity, in terms of
algorithmic descriptions which can be analysed very precisely to establish correct behaviour of function and
timing. The method is manual although it has the potential
for computer assistance.

An attraction of the method is that designs can be
developed and analysed prior to selecting the particular
method of implementation. Once the design is nearing completion, different implementations may permit differing
degrees of optimisation of efficiency and timing behaviour.
But the bulk of the creative design decisions may be made
before any method of implementation is selected. Obviously
if it is known from the start, then design and optimisation
may proceed together, as usually happens with an experienced designer.

Thus the method is applicable to logic designs based
on the following range of approaches:

(a)   SSI and MSI circuit level designs
(b)   LSI programmable component level design (pro-
       grammable logic arrays (PLA), programmable
       array logic (PAL), read only memories (ROM)).
(c)   Semi-custom gate array designs
(d)   Custom design random logic
(e)   Custom design PLA's

Approaches (a), (c) and (d) are very similar although
the latter two currently have a more limited selection of
logical components. Similarly (b) and (e) have a common

characteristic, although there are more degrees of freedom
with custom designed PLA's.

This chapter will be concerned with introducing the
design method while chapter 9 concentrates on implementation
with gates and regular array structures. The material will
concentrate on the concepts rather than the practice but a
simple case study will be cited to support the text.

7.2   THE DESIGN APPROACH

There are many ways of designing a digital hardware
system but all invariably start with an informal written or
verbal description: the ASM method is no different.
Methods based on copying existing designs, or modifying
part of an existing design, or drawing out logic equations
or gate descriptions are not considered. With ASM the
designer is encouraged to think algorithmically to form a
new solution from the requirements stated in the specification. First a set of data path requirements are established and then the control interface specified, together
with the sequence as defined by the algorithm. The controller is described by a flow chart, using a rigid notation, from which a state table is produced. From this
point onwards the process is entirely mechanical producing
programming patterns for regular structures or boolean
equations for gate logic methods. The designer is encouraged to be creative only at the early stages, thereby sep-
arating the arbitrary decision making steps in design from
detailed implementation steps of the later stages. This
sequence will form the structure of this lecture.

7.3   DESIGN EXAMPLE

A case study will be used to provide examples in the
text and therefore the specification of the example is now
stated.

A video game is to be controlled by logic which performs the following functions. (A simplified specification
is given here for clarity).

(i)	Coins of 2p and 10p are recognised by a coin unit.
(ii)	A game costs 10p only.
(iii)	The game is started when the correct money is provided and a start button is pressed.
(iv)	The player is given 4 lives, but can get an extra life for every 500 points obtained in the game.
(v)	The game terminates when all the lives are expended.
(vi)	The controller will indicate, using two lamps driven by logic signals, (a) 'game over' when the game is completed and more money may be inserted, (b) 'start game' when suf- ficient money has been inserted.
(vii)	The game may be activated by a logic signal

and cleared, ready to start again, by a second logic signal. The game will report, by
logic signals, when a life has been lost and
when another 500 points have been obtained.

(viii)   The machine is to be clocked fairly rapidly;
say every 1mS, and all input signals will be
active for a single clock cycle.

(ix)   All outputs must be held in the active or
inactive state all the time the action is
performed. Clearing the game need only be
for 1 clock cycle.

A block diagram of the proposed system is shown in
Fig 7.1. Items (viii) and (ix) above are really design
decisions rather than part of the specification, but they
are necessary to be able to complete the design. Their
relevance will appear as the design proceeds. This specification is probably more detailed than one might normally
expect but there are still many points not stated which
will emerge as the design proceeds.

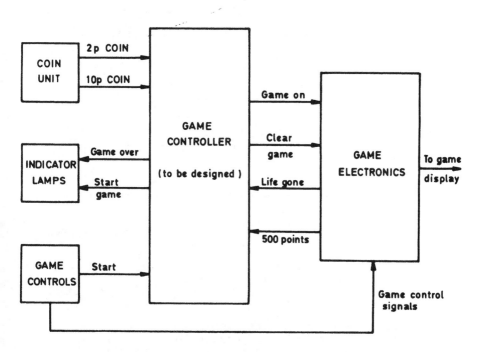

Fig 7.1   Overall view of game controller

7.3.1 Algorithm of Solution

The first step in forming a solution is to establish
an algorithm that produces the required effects. This
phase is intuitive and relies very much on the designers
experience. Frequently the algorithm will not be stated
or written down but will be obvious in the mind of the
designer with experience. However, here a detailed algorithm will be produced to emphasise the process and provide
an example of both the advantages and disadvantages of an
explicitly written algorithm. Fig 7.2 shows a PASCAL
style description of the solution proposed. It should be
obvious that the description contains two interacting
parts; the control part which is concerned with the
sequence of actions and the data part which manipulates
operands using functions. The two sections will now be
described.

```
Startgame:=  false
while true do
    clear game:=  true;  gameover:=  true
    coincount:=  0
    while coin count < 10p do
        game over:=  true
        if 2p detected then coin count:=  coin count +1
           *count in 2p units
        if 10p detected then coin count:=  coin count + 5

    gameover:=  false;  startgame:=  true
    while not start do nothing
    while start do idle      *wait for release
    lives:=  4;  startgame:=  false
    while lives > 0 do
        game on:=  true
        if 500 points then lives:=  lives + 1
        if life gone then lives:=  lives - 1
End of program
```

Fig 7.2 Algorithmic description of solution

7.3.1.1 Data part. Two components of the algorithm are
realised on the data part of the design:

(i)	assignment statements concerned with arith-
	metic and logical operations.
(ii)	relational operations used by conditional
	statements.

Boolean assignments to external or internal units
are performed directly by the controller; for example
GAMEOVER or CLRGAME. External or internal relational variables that are already boolean are connected directly;
for example START or 500POINTS.

Thus
$$coincount := coincount + 1$$
$$and \ coincount < 10p$$

are data part tasks. The objective is to realise the
assignments so that single control lines (instructions)
may invoke the assignment from the control part and the
relational operations so that single boolean variables
(qualifiers) may represent any complex relationship. A
proposed set of data paths is given in Fig 7.3 while
Fig 7.4 shows the mapping between instruction name and
assignment and between qualifier name and relationship.
The design is clearly a simple logical representation
taking no account of implementation constraints. There is
scope for ingenuity in the design irrespective of whether
a logic level design is required (VLSI or semicustom gate
level) or whether specific components are used (MSI or
semicustom functional units). Intuition is still the key
design skill although it should be recognised that a complex data part sub function may well be realised as a
combined data/control part pair which interprets instructions from the higher level.

It is necessary to say a few words here about different instruction types. The PASCAL algorithmic description assumes a well disciplined time behaviour, however with hardware there is more freedom of choice on the
way instructions in fact activate assignment statements
on the data paths. (The full treatment of timing will be
covered later). Three types exist which may be called
delayed, immediate and edge-triggered. All perform the
assignment when active, that is in the true logic state;
this could be logic 0 or logic 1, depending on the logic
conventions used, which is an implementation detail.

delayed ..... performs the assignment on the clock
transition while active.

immediate ... performs the assignment immediately
and so continually forces the assignment while active.

edge-triggered ..... performs the assignment either
on the transition to the
active state or on the transition from the active state.

The delayed instruction is the safest and best approach, but it requires the data path hardware to be designed in a particular way - i.e. synchronous controls
independant of the clock. It is simple to deduce timing
constraints and hazards are no problem. The other two
options require careful design; immediate instructions
are used with asynchronous control inputs - eg clear -
while edge-triggered instructions will be fed to the clock
input of the memory unit. Both will complicate timing calculations; they must be completely glitch free to be safe
but can operate faster than delayed instructions. These
two types of instruction occur frequently in intuitive

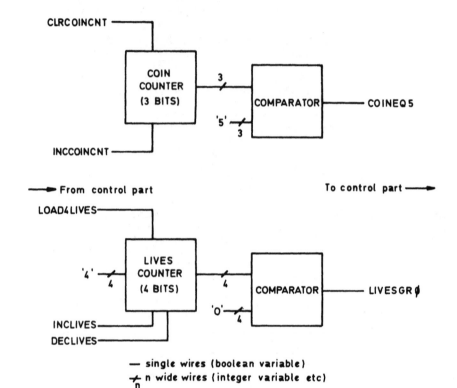

Fig. 7.3   Data paths of game controller

```
Instructions     assignment statements
CLRCOINCNT:      coincnt:= 0
INCCOINCNT:      coincnt:= coincnt + 1
                 (and coincnt:= coincnt + 5 using sequence)
LOAD4LIVES:      lives:= 4
INCLIVES:        lives:= lives + 1
DECLIVES:        lives:= lives - 1

qualifier        relational statement
COINEQ5:         coincnt =< 10p (relies on sequential count-
                                 ing)
LIVESGRØ:        lives > 0
```

Fig. 7.4   Mapping of instructions to assignments and
           qualifiers to relational operations

logic designs. Unfortunately for TTL design and some gate
level semi-custom designs immediate and edge-triggered designs have to be used.

For immediate instructions, the assignment statement
must be changed to distinguish it from a delayed assignment.
An = symbol is used, so for example if the clear instruction of the coin count register is immediate then it would
be described as:

    CLRCOINCNT: coincnt = 0

It is clear that assignments of the form I: = I + 1
become meaningless if expressed in the immediate form since
there is an implied race. The immediate symbol should not
be confused with a relational equals operation which is
described by <> or EQ when used.

For edge-triggered instructions the up arrow and down
arrow may be used to define the action on the rising and
falling edge respectively.

    eg. INC: I↑ = I + 1 or INC: I ∨ = I + 1

The simplicity of the delayed instruction and its
ease of timing calculation makes it far safer to use and
it seems likely that it will be used exclusively in automated design systems. Fortunately VLSI techniques and
some semi-custom techniques are ideally suited to the approach and it is only necessary for the designer to develop an acceptance of the method to fully exploit these
techniques.

The controller can generate composite instructions as
well as single boolean types, for example it may generate
the function code for an ALU unit or a load/shift right/
shift left/ clear for a shift register. Again only the
active states are described.

The qualifier signals are derived from some relational operation on the data paths. They will be two state,
true or false, and can depend on instructions from the control unit - eg. a zero condition from an ALU will depend
on the function being selected by the control unit.

7.3.1.2. Control part. The control part of the design
implements the control sequence of the algorithm using the
well-established model of a finite state machine. The
block diagram is shown in Fig. 7.5 and is composed of two
combinatorial functions and a memory interconnected to
provide a feedback state machine with conditional-state
outputs. Qualifiers come from the data part and any external signals involved in the relational statements: eg. the
START push button. Instructions are fed to the data part
and to any direct external units: eg. the GAMEOVER indicator. The general state machine (GSM) is the most complex
of the five basic machines available. The other four are
shown in Fig. 7.6. The simplest is a combinatorial function with no state, (a), the next, (b), is simply a delay
function with optional combinatorial function. The classic

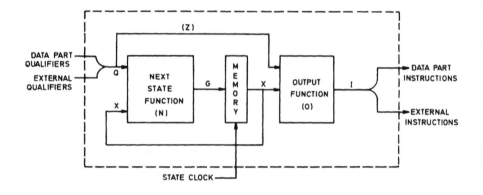

Fig. 7.5.   Model of General State Machine (GSM)

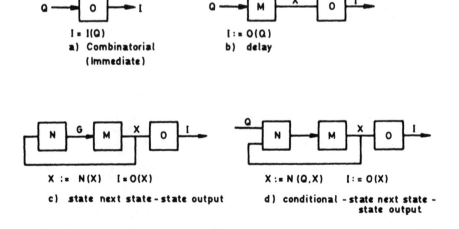

Fig. 7.6.   Restricted versions of the GSM

counter design (with no external controls) is represented
in Fig. 7.6c while the machine with conditional-state next
state function and only state outputs is shown in Fig. 7.6d.
This is probably the most suitable design model since the
full GSM only provides the extra conditional output (see
path Z in Fig. 7.5) which is only used to obtain faster
circuit operation or economise on state variables; i.e.
it is an optimisation step to move from the model of Fig.
7.6d to that of Fig. 7.5. However, our design will be
based on the full GSM. Before discussing how to describe
the algorithm for implementation on the model the timing of
the system will be briefly discussed.

7.3.2 Timing

     Perhaps the most difficult aspect of hardware design
is the determination of timing constraints. The PASCAL
description gives us no clue about timing constraints or
speed of operation of the circuit. A timing analysis is
simply the process of determining whether the setup and
hold time requirements of any memory element are satisfied
when a change occurs in a state variable or an external
signal; the latter will be treated separately.
     If a design is synchronous then all state changes
are derived from a single clock source, some memories being
driven by counter outputs. In this case there can be large
clock skews involved but the analysis can be completed
fairly easily. However, a more rigid definition of synchronous is preferable: every memory element is clocked by
the same clock signal. Now clock skew is a minimum but
the designer must allow for the memory to be updated every
cycle or a clock enable control must be introduced.
     In general, the timing analysis of an asynchronous
design cannot be completed and when it can it takes considerable time. The problem of asynchronous signals is
discussed under the section on external signals.
     All the timing paths of a simple control/data unit
are shown in Fig. 7.7. The simple ones are t4, the control
part feedback loop, t3, the data part memory to control
part memory (qualifier only), and t2, control part memory
to data part memory (delayed state assignment only). The
difficult timing paths are those associated with t1. This
shows a path due to an instruction immediately affecting
a qualifier which must propagate back to the controller
memory. The dotted route, t1I, illustrates the path when
a memory is immediately updated (eg. asynchronous clear)
and this affects a qualifier that affects the next state
function. This route is frequently forgotten in an analysis especially by inexperienced designers. If the data
part contains a bypass path, like Y, then the use of the
GSM can produce particular problems since a combinatorial
loop can be generated via path Z in the control part, C2,
C3, path Y, C4 and back to path Z. Extreme care is needed
to ensure that this cycle is not activated so that a race
occurs.

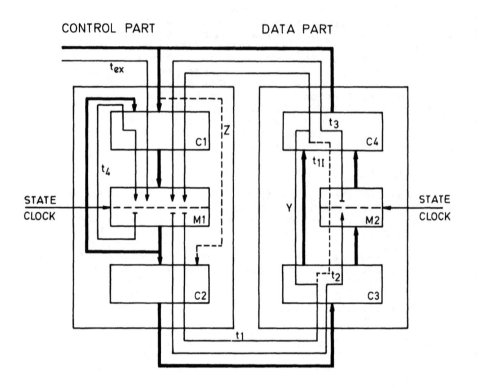

$$t_1 = t_{DM1} + t_{DC2} + t_{DC3} + < t_{DM2} > + t_{DC4} + t_{DC1}$$

<> optional term

$$t_2 = t_{DM1} + t_{DC2} + t_{DC3} + t_{SM2}$$ t_{DX} delay of X

$$t_3 = t_{DM2} + t_{DC4} + t_{DC1} + t_{SM1}$$ t_{SX} set up of X

$$t_4 = t_{DM1} + t_{DC1} + t_{SM1}$$ t_{ex} external time

Fig. 7.7 Timing cycles of design model

Clearly if all the instructions are derived from the state directly (no conditional instructions) and they drive delayed assignments on the data path, then timing analysis becomes relatively simple, especially if the memories are all clocked directly by a single clock. With such design restrictions automated timing analysis is feasible. The UK5000 ULA system design style is based on this approach and timing analysis is performed entirely automatically.

7.3.3 External signals

Analysing the timing of external signals is only possible if they are synchronous, in which case they become an extension of the data part analysis. If they are asynchronous then guaranteeing the correct operation is only possible by synchronising them. Unsynchronised signals fed into a state machine will only cause random spurious failures since set-up times cannot be guaranteed. If external asynchronous signals are fed to the data paths then the designer is inviting disaster.

7.3.4 Controller description

The controller is described by an algorithmic state machine (ASM) chart (Clare (1)). The design for the game controller is shown in Fig. 7.8. It shows the sequence of instructions generated, according to the state of the qualifiers, as the state clock advances. Instructions are contained in state boxes and conditional output boxes, which indicates the state of state/condition respectively in which they are active. Inactive instructions are not shown. The qualifiers are labelled inside condition boxes and indicate the path of the state transition when the qualifier is either true or false. Every link path (route from a state) must terminate on the same or another state box. Cyclic link paths are invalid. Equally, paths that lead to two or more next states simultaneously are invalid. Each state box is labelled with a state name to identify it.
Unlike other methods, only qualifiers that influence the current state transition are described. For example in state 5, START has no effect and so is not described. Thus although the controller has a total of 7 qualifier inputs this does not result in 128 paths being shown on the ASM chart for every state. In fact there are only a total of 16 link paths in all; this is the maximum of product terms needed in the implementation.
The structure of the ASM chart follows directly from the algorithmic description except for a few optimising steps which arise because of the mismatch between the interpretation of the assignment statements in a program and the hardware realisation of them. A new state is not necessarily required for every assignment statement, and states may not be necessary to implement the relational statements; removal of these is an optimisation process and they could be inserted initially and then removed. This is because a hardware implementation can perform many assignments sim-

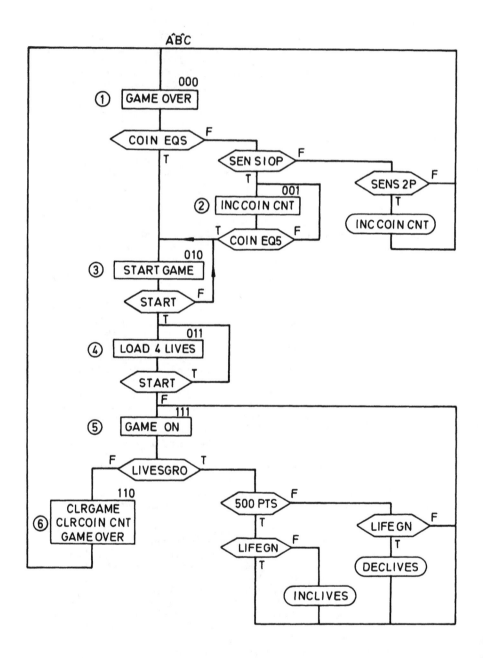

Fig. 7.8.   ASM chart of the Video Game Controller

ultaneously (see state 6) and evaluate one or more relational statements and external qualifiers simultaneously. Thus
an action and a test can be performed together; eg. in state
5 LIVESGRO, 500 points, LIFEGN are being tested continuously while INCLIVES and DECLIVES are activated when required.
Notice, though, that because DECLIVES is delayed, it is not
possible for LIVESGRO to change state in the same state time
that DECLIVES is activated, so no potential race or conflict
exists.

Little more can be said about the creation of the ASM
chart, it is a creative design process and therefore is very
much a personal issue. However, it is important to notice
how the method can cope with complex design problems. This
design has 6 states, 7 inputs (qualifiers) and 9 outputs
(instructions) which represents a modestly complex design
problem. Designs with 16 states, 18 inputs and 14 outputs
have been easily handled and implemented with a mixture of
PLA's, multiplexers and random logic. The description is
also easy to interpret. Another designer could easily
understand how it works and what was in the original designers mind. Thus it forms an invaluable documentation aid.
The method is described very fully in reference (1), although the notation has been modified and the algorithm to data
part/control mapping is more fully described here.

The description produced so far has taken no account
of implementation, and has not had to; this is one of its
attractions. To implement the design the description needs
a state assignment and then transformation to a state table
description.

7.3.5 State assignment

This process is common to many other methods of logic
design and the same ideas and rules apply. The first process involves selecting the number of state variables according to the following inequality:

$$2**n >= N$$

where n is the number of state variables (bits in the memory
register) and N the number of states in the design. Selecting n larger than the minimum may lead to simpler logic or
be more appropriate for the implementation used. For
example in a PAL simple logic functions with many state
variables are provided, whereas in a PLA complex logic is
easily implemented but outputs may be limited.

Once the number of variables has been decided then
each state in the design must be assigned a state code. A
possible assignment for the video game controller is:

State name	State code	State name	State code
1	000	5	111
2	001	6	110
3	010		
4	011		

There has been much work to find optimum state assignments particularly with a view to reducing the number of
logic gates in the design. This may be important with random logic designs, semi-custom gate designs and possibly
with VLSI random logic structures. With LSI programmable
components like PLA's, PAL's or PROM's containing a fixed
function it is often a fruitless exercise spending too much
time on optimising. It is necessary to ensure the design
fits the device which usually involves far simpler issues
like pin count or number of product terms. For example the
maximum number of product terms for a PLA implementing an
ASM chart directly is the number of link paths on the chart;
state assignment can reduce this a small amount if link
paths with all outputs inactive can be assigned.

State assignment is critical if asynchronous inputs
are included. Here the objective will be to keep the two
next states, associated with the two states of the asynchronous variable, one bit distance apart. Thus states 3
and 4 and states 4 and 5 should be one bit distance apart
to prevent the asynchronous input, START, causing an invalid
state transition.

Notice that it is only possible to satisfy this requirement with a single asynchronous variable in any ASM
block. (An ASM block is the sub-chart containing a single
state box and all the other boxes up to but excluding all
the next-state boxes). If the block is complex, having
many next states, then it may still be impossible to make a
safe state assignment; hence the necessity of avoiding asynchronous inputs.

Once state assignment has been completed then the
state table can be produced. This will form the starting
point of chapter 9.

7.4 SUMMARY

This chapter has described how a more formal approach
to logic design can be adopted that uses an algorithm of a
solution as a basis for designing a hardware logic system.
The method partitions the design into data part and control
part and the paper shows the relationship between these
parts and the algorithm. The data part implements assignment and relational statements, while the control part
activates the former and responds to the latter in the correct sequence. Optimisation was shown to form an integral
part of this process. An ASM chart is produced to describe
the controller and this chart forms the basis of the implementation decisions that follow. These will be the subject
of chapter 9. The method can be manual but is also suited
to computer aided processes especially if some constraints
are imposed on the implementation.

REFERENCES

1.   Clare, C. "Designing logic systems using state
             Machines". McGraw Hill, 1973.

2.   Winkel, D., Prosser, F. "The art of digital Design".
             Prentice Hall, 1980.

3.   Fletcher, W.I. "An engineering approach to digital
             Design". Prentice Hall, 1980.

4.   Peatman, J.B. "Digital hardware design". McGraw
             Hill, 1980.

Chapter 8

The programmable logic array: implementation and methodology

D.J. Kinniment

8.1   INTRODUCTION

There are a number of methods for implementation of a semi-custom design as an integrated circuit, but the main thrust of most methods is to choose a target structure on the silicon which is both flexible in use and permits a simple, correct compilation process from the description of the required behaviour to the mask layout.

The emphasis is therefore that of quick implementation with some inevitable sacrifice of performance and silicon area.   As the number of variables allowed in the target structure reduces, the compilation process can be made easier but the range of application for which the structure is efficient usually becomes more restricted.   Uncommitted Logic Arrays (ULA's) only permit a small set of primitive circuits (usually gates) and only allow their positioning at discrete points on a regular array.   They are therefore inefficient at implementation of very specialised systems such as memory, but otherwise fairly flexible in their range of application.   On the other hand they may suffer from a degree of difficulty in layout because of the fixed amount of space allowed for interconnections.   Programmable Logic Arrays (Carr and Mize (1)) substitute an overall floor plan, or architectural arrangement of the gates for the fixed array of primitives; the structure is therefore simple to connect because the signal flow is predefined by the architecture, and can be more efficient in silicon area because only the area actually required for the function is used rather than a fixed size of chip.   The disadvantage of the PLA is that its fixed architecture narrows the range of applicability more than other semi-custom techniques.

8.2   IMPLEMENTATION

Since any binary function can be expressed as a sum of minterms, it is possible to construct any function of one output as a sum of products circuit (Fig.8.1).   This well known method can also be applied to multiple output functions, in which each output is formed from the OR of its own set of product terms as shown in Fig.8.2.   In this diagram each product term is formed by a separate AND gate, but there is a significant likelihood that the same product

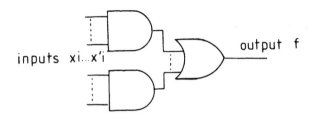

Fig.8.1 Sum of products circuit for a single output

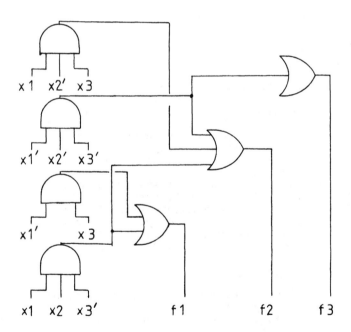

Fig.8.2 Multiple output function

term will be needed in two or more of the required output functions. It is thus possible to use the output of the AND gate producing that term in each of the summing OR gates rather than duplicate it.

The PLA is a structure using this method in a generalised way as shown in Fig.8.3. Here one AND gate exists for
each unique product term, and one OR gate for each output.
     If n inputs are required to form o outputs, and a total
of m unique product terms appear in the functions representing the outputs, each of m AND gates must be capable of
accepting up to 2n inputs, and each of o OR gates must be
capable of accepting up to m inputs.

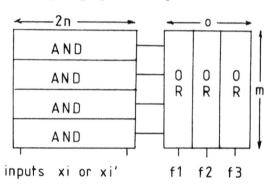

Fig.8.3 P.L.A. Structure

     Usually the actual layout of the PLA also follows the
diagram of Fig.8.3 and will consist of an m x 2n AND plane
followed by an m x o OR plane, in which the actual functions
implemented depend on which inputs are connected to each AND
gate, and which product terms are connected to each OR gate.
This relatively simple interconnection scheme is usually
made even simpler by arranging that every input is made
available to every AND gate, and every product term is available to every OR gate.
     Programming the array therefore reduces to specifying
a) the number of inputs (n), outputs (o), and product terms
(m); and b) two Boolean matrices, one 2n x m indicating that
a particular input (or its complement) is used in each AND
gate, and the other m x o indicating that a particular
product term is used in each OR gate.

8.2.1 NMOS Layout

     In any particular technology truly complementary struct-
ures are usually difficult to achieve, and one version often
has advantages over the others performance even if both are
possible.
     This is the case with NMOS technology where NAND gates
can be made as well as NOR; however, in terms of performance
and ease of design NOR gates have an advantage (assuming a
high voltage level represents true). This is mainly because
NAND gates require an increase in the length/width ratio of
the pull up for every extra input, and consequently decrease
in speed as the number of inputs increases. This effect can
be avoided by decreasing the length/width ratio of the pull

downs instead, but it is usually considered more convenient to use a NOR-NOR implementation of the PLA as the geometry of the gates is not dependent on the function. The example of Fig.8.2 is shown as a NOR-NOR NMOS PLA circuit in Fig.8.4. In an NMOS silicon-gate process the NOR gate input connections are normally made with polysilicon lines and the pull-down transistor drains are joined with metal to form the output of the gate. Programmability is usually achieved by placing an n-channel transistor connected between ground and the gate output under the appropriate input line, though it is possible to use other methods such as metal links to join an existing transistor.

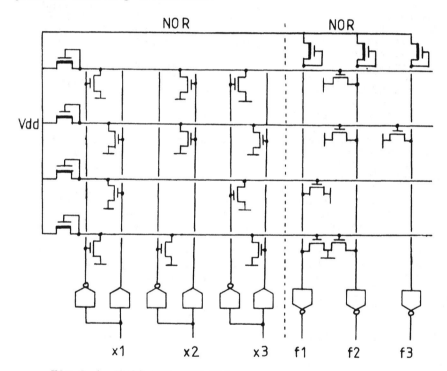

Fig.**8**.4   NMOS NOR-NOR PLA

8.2.2   CMOS Layout

In a completely static CMOS version of the NMOS PLA each of the depletion-mode pull-up transistors would have to be replaced by a series chain of p-channel transistors with one p-channel device for each n-channel in the pull-down section. Absence of an n-channel would be complemented by short circuiting the corresponding p-channel device. This method suffers from two difficulties; firstly a doubling of the area required compared with the NMOS version and second-ly the inconvenience of the variability in the length/width ratio required of the series transistors.

If a clock exists in the system it is usually possible to combine the low power dissipation of CMOS with the small area of NMOS by using dynamic p-channel pull-ups which are switched on to precharge the output lines when any output change is required.

8.2.3 The PLA as a Structure

A multiple output binary function can be minimised and laid out either as a ULA or a PLA, but there are a number of advantages to the structured layout inherent in the floor plan of the PLA. These are mainly concerned with the separation of the layout from the behaviour. Given the required behaviour in terms of Boolean equations it is possible to estimate the number of product terms required, and from that the exact area and disposition of inputs and outputs follow relatively simply.

Similarly it is not necessary to specify a function precisely in order to have a close estimate of the topology and area of the PLA required to implement that function. Top-down design thus becomes easier using PLA's and detailed modifications of the function can be done at a late stage without extensive rework.

Another important aspect of the regular layout is that inputs and outputs can be permuted in their order simply by altering the columns of the AND or OR plane matrices; there is no significance to any particular column when compared to another.   This may be of advantage in interfacing the structure to some other, less modifiable, structure in minimising the connection area.   In Fig. 8.4 the inputs and outputs are shown at the bottom of the planes. It is possible to put them either at the top without loss of generality, or at the sides with some loss.   It is this flexibility combined with the ability to postpone the functional binding to a later stage that makes the PLA a useful structure.

8.3 COST PERFORMANCE IMPROVEMENT

A number of useful functions have characteristics which make them unsuitable for implementation as a simple PLA. An example is that of addition, where the number of product terms needed for a one-bit full adder, to produce the sum and carry is 7. As the number of inputs, n, increases the number of product terms is related to 2^n so that a two-bit full adder requires 23, a 3-bit adder over 100, etc. Since the silicon area is directly related to the number of product terms, a 16-bit adder is just not feasible.

As well as straightforward Boolean minimisation of the functions other techniques can be used which can make the area considerably smaller.

These include
- (a)   The use of decoders on the inputs
- (b)   The use of exclusive or functions on the outputs

and (c)   Topological minimisation (folding) to reduce unused areas.

8.3.1 Multilevel PLA's

Weinburger (2), describes a number of designs in which the methods given above have been used to give a reasonable size for 8, 16 and even 32-bit adders implemented as a PLA.

The number of product terms required can be reduced partly because the addition function is symmetric; that is both A and B inputs have equal weight in producing the sum S, but mainly by using more than two levels of logic. An input decoder consisting of four NOR gates is used at each bit position. This takes the A and B inputs and produces four outputs one of which is true for each possible value of A and B, and carries exactly the same information in the same number of wires as the true and false values of A and B used in a normal PLA. It is possible now to produce the exclusive OR function A XOR B with only one product term rather than two since A XOR B = (A NOR B) NOR (A' NOR B').

Using XOR gates on the outputs in an adder also produces quite large savings since in any bit position sum = A XOR B XOR carry. Each sum output can thus be broken down into two functions having far fewer product terms than the single equivalent function. Fig.8.5 shows how an 8-bit adder can be reduced to 25 product terms when NOR input decoders are used to produce the signals, $D_0=A'.B'$, $D_1=A'.B$, $D_2=A.B'$, $D_3=A.B$, and XOR gates are available at the

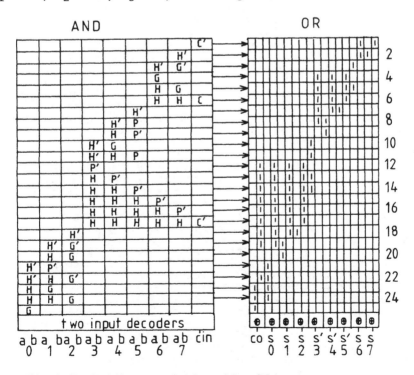

Fig.8.5 Weinburger 8-bit adder PLA

outputs. In the figure the product terms are made by NOR gates realising the following functions:-

$P = A+B = D_0'$
$G = A.B = (D_0+D_1+D_2)'$
$H = A \oplus B = (D_0+D_3)'$
$P' = (D_1+D_2+D_3)'$
$G' = D_3'$
$H' = (D_1+D_2)'$

This has mainly been achieved by decomposing the NOR-NOR PLA to a NOR-NOR-NOR-XOR structure and a more general investigation of three-level, or NOR-NOR-NOR PLA's has been carried out by Sasao (3) in which mode upper bounds on the size of such PLA's is given for some cases.

8.3.2 Folding

More than half of the PLA shown in Fig.8.5 is not active and does not usefully contribute to the function. It is possible to recover some of this lost area by topological manipulation of the array, for example by noting that product terms 18 to 25 do not have any inputs in common with product terms 1 to 8. They could therefore equally well occupy the unused area on the top left of the PLA with the outputs appearing on the left hand side rather than the right hand side. If product terms 9 to 11 appear on the right hand side, and 12 to 17 on the left hand side C_0, S_0, S_1, and S_2 can be completely formed in an OR plane on the left whilst S_3', S_4', S_5', S_6, and S_7 are formed on the right.
The height of the PLA is now reduced from 25 product terms to 17, and a small saving in the width can also be made by sharing an OR plane column between S_3' and S_7 as shown in Fig.8.6.
This kind of manipulation is known as folding, but it has the disadvantage of reducing the permutability of the inputs and outputs at the same time as reducing the area.

8.3.3 Performance

In large PLA's the capacitive loads represented by the long lines to all gates in the AND plane and from the product term outputs to all gates in the OR plane limit the speed of response. Thus the larger the array the worse its performance becomes in direct proportion to the number of inputs and outputs or the number of product terms. This effect can be offset to some extent by increasing the drive capability of the input drivers, but since the input lines are usually polysilicon their series resistance coupled with their capacitance provides an inherent delay proportional to $(length)^2$. Very large PLA's are therefore impractical because of both large area and consequent poor performance.
A way of dealing with large complex functions is to decompose them into a number of smaller functions, each of which may be easily implemented as a PLA with relatively high performance. As shown in section 8.3.1 decomposing a single two-level PLA effectively into two two-level PLA's gives considerable area savings, and the resulting system

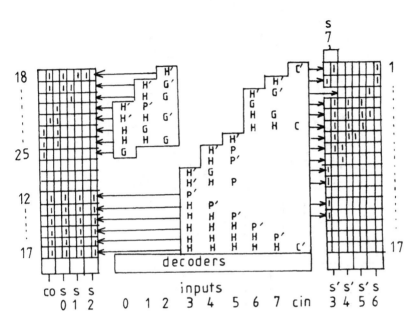

Fig.8.6   Folded adder PLA

will be much faster than the original.
     An example of the technique of decomposition is shown
in Fig.8.7 in which a 4-bit x 4-bit multiplier is constructed
from four 2 x 2-bit multiplier PLA's and nine single-bit
adder PLA's. Here the two input operands are each split into
a most significant half and a least significant half. Four
partial products are then formed in the 2 x 2 multipliers
and the results added in two cascaded rows of adders.
     Clearly the process of decomposition can be carried to
the point where the individual PLA's are reduced in function
to something like, 2 input-gates and the structure closely
resembles a ULA.
     It appears likely, however, that there is an optimum
size of PLA in terms of number of inputs and outputs and
number of product terms which will give the best functionality for a given area-time product and that this is a more
complex unit than the simple gates used in today's ULA's.

8.4   FINITE STATE MACHINES

     Controllers and microprogrammed systems (Mead & Conway
(4)) are examples of the use of finite state machines in
which the outputs and the next state are a function of the
inputs and the present state. Any finite state machine can
therefore be implemented with a register to hold the values
and the present state, a multiple-output combinational

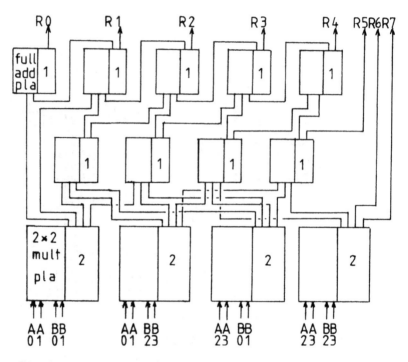

Fig.8.7   Multiplier decomposition into PLA's

circuit to evaluate the outputs and the next state from
these, and a second register to hold their values after
evaluation.

The system is completed by passing the next state to
the present state register when the cycle is complete as
shown in Fig.8.8 where a PLA is used as the combinational
circuit.

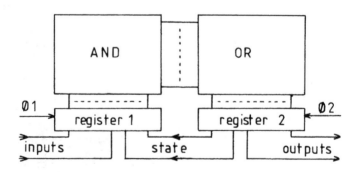

Fig.8.8   Finite state machine

The two clocked registers are particularly easy to produce in NMOS technology where they can be dynamic storage nodes charged or discharged through a pass transistor.

The process of design for a finite state machine consists of drawing up a state table in which all the possible outcomes of the next clock period are listed under each state; that is, the next state and the resulting outputs for each allowed input.

This is usually first done in the form of a state diagram, and a simple example of a 4 state machine is shown in Fig.8.9

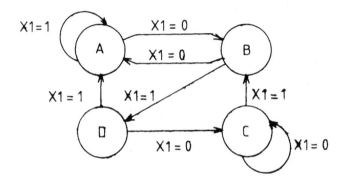

Fig.8.9   State diagram

In this example the machine has only one input, X1, and one output which is not shown on the diagram but indicates that state A has been reached followed by the input X1 becoming zero.

The state table which follows from this diagram is a formal specification of the behaviour required, and if it is written in a machine readable form can be used as the first step in a series of processes resulting ultimately in the fabrication of a controller chip having the specified behaviour from automatically produced masks.

Each of the processes required to produce the masks is relatively simple and easy to automate because the form of the specification at each stage is well defined and the translations required between one format and the next are well known.

Fig.8.10 shows a machine readable state table definition produced by hand from the state diagram. The meaning of the table is fairly self evident. A program, Morris (5), reads and checks the syntax of the text, assigns codes to the states and outputs a set of Boolean equations defining the PLA. Other software Lim (6) minimises the number of product terms required and outputs the PLA structure required. The results of these steps can be seen in Figs.8.11 and 8.12.

```
CONSTANT
   Set=1
   Clear=0
INPUTS
   x1
OUTPUTS
   Flag

TABLE

STATE A
   IF x1 GOTO A DO Flag=Clear;
   IF NOT x1 GOTO B DO Flag=Set;

STATE B
   IF x1 GOTO D DO Flag=Clear;
   IF NOT x1 GOTO A DO Flag=Clear;

STATE C
   IF x1 GOTO B DO Flag=Clear;
   IF NOT x1 GOTO C DO Flag=Clear;

STATE D
   IF x1 GOTO A DO Flag=Clear;
   IF NOT x1 GOTO C DO Flag=Clear;
```

Fig.8.10   State table description

```
INPUTS: INO, FEED2, FEED1;      (x1, x2, x3)

INO.FEED1'.FEED2+INO'.FEED1.FEED2'+INO'.FEED1.FEED2;
INO'.FEED1'.FEED2'+INO.FEED1'.FEED2+INO.FEED1.FEED2';
INO'.FEED1'.FEED2';
:


AFTER MULTIPLE OUTPUT MINIMISATION
THE NUMBER OF INPUT VARIABLES ARE          3
THE NUMBER OF OUTPUT VARIABLES ARE         3
AN EXTRA OUTPUT COLUMN HAS BEEN ADDED TO MAKE THE TOTAL
OUTPUT COLUMN EVEN
THE NUMBER OF PRODUCT TERMS ARE            4

THIS IS THE ANDPLANE ARRAY
   INO      FEED2    FEED1
   1 0      1 0      0 1          ⎡ x1  x2 x3'  ⎤
   0 1      0 0      1 0          ⎢ x1'     x3  ⎥
   0 1      0 1      0 1          ⎢ x1' x2'x3'  ⎥
   1 0      0 1      1 0          ⎣ x1  x2'x3   ⎦
THIS IS THE ORPLANE ARRAY
   1 2 3 4
   1 1 0 0
   1 0 0 0
   0 1 1 0
   0 1 0 0
```

Fig.8.11   Resulting equations and minimisation

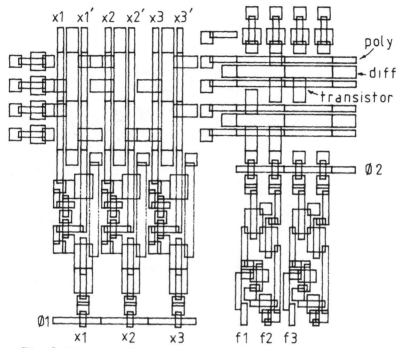

Fig.8.12 Polysilicon and diffusion layers of
PLA showing programming transistors

8.5 CONCLUSION

For a 5μ geometry i.c a typical PLA cell size is 40μ x
40μ, where the number of cells required to implement a PLA
is (2u + o) x m. In the adder of Fig.8.6 (32 + 16) x 17
cells are required and the approximate area would be 1.96mm
x 0.68mm ≃ 1 1/3mm². An 8-bit adder implemented on a ULA
would require of the order of 200 2-input cells, each more
complex than the single transistor PLA cell and typically
100μ x 100μ taking into account the need for interconnection
space. The area is therefore about 2mm² for the same
function, and this demonstrates that a PLA can be a very
efficient methodology for a range of simple functions.

It cannot be used, however, to provide memory, apart
from its obvious use as a lookup table or ROM so that its
range is more restricted than ULA which can provide a
limited amount of register storage, though not as efficient-
ly as a custom design.

The advantage of the PLA is in its simple compilation
which allows a very fast and correct turn around time from
behavioural description either as a set of binary functions
or a state table to the layout itself.

REFERENCES

1. Carr, W.N., and Mize, J.P., 1972, 'MOS/LSI Design and Application', McGraw Hill Book Co., Inc., New York.

2. Weinburger, A.P., 1979, 'Highspeed Programmable Logic Array Adders', <u>I.B.M. J. Res. Develop</u>, <u>23</u> <u>2</u>, 163-179.

3. Sasao, T., 1981, 'Multiple-Valued Decomposition of Generalised Boolean Functions and the Complexity of Programmable Logic Arrays', <u>IEE Trans</u>, <u>C-30</u>, <u>9</u>, 635-643.

4. Mead, C., and Conway, L., 1980, 'Introduction to VLSI Systems', Addison Wesley, New York.

5. Morris, D., 1982, 'The STATIC program Manual', Final Year Student Thesis, Computer Science Department, Leeds University.

6. Lim, C.Y., 1982, 'A PLA minimiser program', Final Year Student Thesis, Electrical and Electronic Engineering Department, University of Newcastle upon Tyne.

Chapter 9

PLA and ROM based design

E.L. Dagless

9.1   INTRODUCTION

The ASM method of design was briefly introduced in
chapter 7 and it was shown how an abstract design description is developed from an algorithmic statement of a solution.   A data part is designed intuitively and a control
part described by means of an algorithmic state machine
chart.   The process of state assignment is the first step
in the implementation of the controller.   In this section
the full range of implementation options is discussed, con-
centrating particularly on methods using programmable LSI
components and semi-custom and VLSI implementations.   The
three approaches are:

        (i)   gates and multiplexers,
        (ii)   PLA's and PAL's,
        (iii)   read only memories (ROM).

Gate methods cover conventional random logic using
TTL and CMOS integrated circuits as well as gate level
methods of semi-custom and VLSI design.   Multiplexer based
solutions suit standard MSI integrated circuit methods and
semi-custom and VLSI methods where a regular structure is
desirable but PLA's or ROM are not appropriate.   PLA and
PAL methods are suited to modern programmable LSI circuits
based on these structures and also to VLSI designs which
incorporate a PLA structure.   ROM based methods are princi-
pally intended to exploit LSI and VLSI ROM and PROM integrated circuit components since in custom VLSI designs,
PLA's offer a more attractive approach.

However, irrespective of the implementation method
chosen the ASM technique is suited to all, and the design
process so far should not be affected.   In practice the
experienced designer may well have optimised parts of
the design solution if the implementation is known when
producing the data part and the control part descriptions.
All three methods require the production of a state table
from the ASM chart.

9.2   STATE TABLE

The state table is a tabular representation of the

ASM chart describing in a truth table like form, the two
combinatorial functions of the general state machine; the
next state function, N, and the output function, O. The
table for the design of the video game controller is given
in Fig. 9.1. The symbolic form of the table could be produced before any state assignment is performed, and the
logic-level table shows the situation if all the signals
have a positive logic convention; thus some knowledge of
polarity of signals must be known to produce Fig. 9.1b.

The table shows the two groups of inputs at the top ;
qualifiers, Q, and state, X, while the bottom half
shows the outputs of the two combinatorial blocks; next
state, G, and instructions, I. The hyphens on the input
side are don't care input terms while the decimal points on
the output are inactive levels. Notice that the number of
lines in the table is equal to the total number of link
paths on the ASM chart; therefore an assessment of the logical complexity can be obtained before state tables are
produced.

The symbolic state table contains no more information
than the ASM chart, it is simply a more convenient representation from which to start the implementation phase.
The signal-level table reveals more implementation detail
concerning the active level assignment of signals. When
testing a design the ASM chart is the more convenient form
of description to follow, therefore, logic-level assignments should be shown in mnemonic form in the signal name,
on the ASM chart, for future use. This can be done by pre-
ceeding the name with an N, for negative logic, or L for
active low or some other convention. It is also useful at
this stage to indicate immediate or edge-triggered instructions by appropriate naming conventions.

Thus, if the assignment process to clear the coin
count is in fact active low and an immediate instruction
then the mnemonic could become: ILCLRCOINCNT. All the
implementations to be discussed start from either Fig. 9.1a
or Fig. 9.1b. In an automated design process the state
table may not be visible at all but be merely retained in
the computer as an internal data structure (e.g. Edwards
et al (2)).

9.3   GATE BASED DESIGN

In this category both random logic methods and multiplexer based methods will be discussed. Both are applicable to conventional SSI and MSI techniques and they may
also be applied to semi-custom methods and VLSI (although
not the preferred approach for the latter). The starting
point for both methods is the logic-level description of
the state table in Fig 9.1b, and the method is the classi-
cal approach of using Karnaugh Maps followed by Boolean
Algebra.

9.3.1   Random logic

For each output variable a Karnaugh map (K-map) is

QUALIFIERS (Q)																	
COINEQ5	T	F	F	F	T	F	-	-	-	-	-	-	-	-	-	-	I
SENS1OP	-	T	F	F	-	-	-	-	-	-	-	-	-	-	-	-	N
SENS2P	-	-	T	F	-	-	-	-	-	-	-	-	-	-	-	-	P
START	-	-	-	-	-	-	T	F	T	F	-	-	-	-	-	-	U
LIVESGRO	-	-	-	-	-	-	-	-	-	-	F	T	T	T	T	-	T
500PTS	-	-	-	-	-	-	-	-	-	-	-	T	T	F	F	-	S
LIFEGN	-	-	-	-	-	-	-	-	-	-	-	T	F	F	T	-	
STATE (X)																	
NAME	1	1	1	1	2	2	3	3	4	4	5	5	5	5	5	6	

NEXT STATE (G)																	
NAME	3	2	1	1	3	2	4	3	4	5	6	5	5	5	5	1	O
INSTRUCTIONS (I)																	
GAMEOVER	A	A	A	A	.	.	.	.	.	.	.	.	.	.	.	A	U
INCCOINCNT	.	.	A	.	A	A	.	.	.	.	.	.	.	.	.	.	T
STARTGAME	.	.	.	.	.	.	A	A	.	.	.	.	.	.	.	.	P
LOAD4LIVES	.	.	.	.	.	.	.	.	A	A	.	.	.	.	.	.	U
GAMEON	.	.	.	.	.	.	.	.	.	.	A	A	A	A	A	.	T
CLRGAME	.	.	.	.	.	.	.	.	.	.	.	.	.	.	.	A	S
CLRCOINCNT	.	.	.	.	.	.	.	.	.	.	.	.	.	.	.	A	
DECLIVES	.	.	.	.	.	.	.	.	.	.	.	.	.	.	A	.	
INCLIVES	.	.	.	.	.	.	.	.	.	.	.	.	.	A	.	.	

Fig. 9.1a   Symbolic state table

QUALIFIERS (Q)																	
COINEQ5	1	0	0	0	1	0	-	-	-	-	-	-	-	-	-	-	I
SENS1OP	-	1	0	0	-	-	-	-	-	-	-	-	-	-	-	-	N
SENS2P	-	-	1	0	-	-	-	-	-	-	-	-	-	-	-	-	P
START	-	-	-	-	-	-	1	0	1	0	-	-	-	-	-	-	U
LIVESGRO	-	-	-	-	-	-	-	-	-	-	0	1	1	1	1	-	T
500PTS	-	-	-	-	-	-	-	-	-	-	-	1	1	0	0	-	S
LIFEGN	-	-	-	-	-	-	-	-	-	-	-	1	0	0	1	-	
STATES (X)																	
A	0	0	0	0	0	0	0	0	0	0	0	1	1	1	1	1	
B	0	0	0	0	0	0	1	1	1	1	1	1	1	1	1	1	
C	0	0	0	0	1	1	0	0	1	1	1	1	1	1	1	0	

NEXT STATE (G)																	
NA	0	0	0	0	0	0	0	0	0	0	1	1	1	1	1	0	
NB	1	0	0	0	1	0	1	1	1	1	1	1	1	1	1	0	O
NC	0	1	0	0	0	1	1	0	1	1	0	1	1	1	1	0	U
INSTRUCTIONS (I)																	T
GAMEOVER	1	1	1	1	.	.	.	.	.	.	.	.	.	.	.	1	P
INCCOINCNT	.	.	1	.	1	1	.	.	.	.	.	.	.	.	.	.	U
STARTGAME	.	.	.	.	.	.	1	1	.	.	.	.	.	.	.	.	T
LOAD4LIVES	.	.	.	.	.	.	.	.	1	1	.	.	.	.	.	.	S
GAMEON	.	.	.	.	.	.	.	.	.	.	1	1	1	1	1	.	
CLRGAME	.	.	.	.	.	.	.	.	.	.	.	.	.	.	.	1	
CLRCOINCNT	.	.	.	.	.	.	.	.	.	.	.	.	.	.	.	1	
DECLIVES	.	.	.	.	.	.	.	.	.	.	.	.	.	.	1	.	
INCLIVES	.	.	.	.	.	.	.	.	.	.	.	.	.	1	.	.	

Fig. 9.1b   Logic level state table (assumes all signals active high)

produced.   The problem is coping with the large number of
input variables.   The solution is to produce maps with the
state variables as the map variables and then to enter the
qualifiers on the K-maps as map-entered variables.   Thus
the mapping process breaks down into solving each sub-table
for each state in the state table and entering the result
in the square on the map for the state concerned.   The map
for NA, NB and NC is shown in Fig. 9.2.

NC

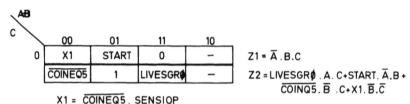

C \ AB	00	01	11	10
0	X1	START	0	—
	$\overline{COINE05}$	1	LIVESGRØ	—

$Z1 = \overline{A}.B.C$

$Z2 = LIVESGRØ.A.C + START.\overline{A}.B + \overline{COINQ5}.\overline{B}.C + X1.\overline{B}.\overline{C}$

$X1 = \overline{COINEQ5}.SENSIOP$

$NC = \overline{A}.B.C + LIVESGRØ.A.C + START.\overline{A}.B + \overline{COINEQ5}.\overline{B}.C + \overline{COINEQ5}.SENSIOP.\overline{B}.\overline{C}$

NB

COINEQ5	1	0	—
COINEQ5	1	1	—

$NB = B.C + \overline{A}.B + COINEQ5.\overline{A}$

NA

0	0	0	—
0	$\overline{START}$	1	—

$\overline{NA} = \overline{C} + \overline{B} + START.\overline{A}$

Fig. 9.2.   Karnaugh maps of next state variables
            NA, NB, NC

To reduce a map with variables in it the process is
similar to the normal method except an extra stage is introduced.   The method is as follows:

    (a)   Make all variables zero and reduce on 1's to
            produce a partial function, Z1.

    (b)   Make all the 1's now used into don't cares
            and reduce on each variable, grouping only
            with the same variable, to produce a second
            partial function of the form:

Z2= var1.P(abc)+ var2.P(abc) + etc.

where var1 and var2 are map-entered variables
and P(abc) is a product term of the mapping
variables, abc.
(c)   Combine the two partial terms, i.e.

Z= Z1+Z2, where Z is the desired result.

The reduction can be carried out on 0's rather than
ones if desired but remember the variable must be complemented to make it zero. From this process the boolean
equations and thence the logic diagram may be produced.
Any further reductions may be made if desired. For example
the equation for the NC signal is:

NC= A'.B.C+LIVESGRO.A.C+START.A'.B+COINEQ5'.B'.C
    +COINEQ5'.SENS1OP.B'C'

Any immediate or edge-triggered instructions must be
examined carefully for 'glitch' free operation; remembering
that as well as having problems with delays through gates,
outputs from the state register are not guaranteed to
change together so there may be a race hazard even with
simple combinatorial functions of the state variables.
The best solution is to ensure a state assignment that
makes the sensitive instruction a direct state output.
This usually requires many more state variables.
      The only problem left is to establish whether the
timing is correct. Obviously the experienced designer will
have made an approximate timing analysis at the ASM chart
level, but a detailed calculation can only be made when
the logic structure has been specified. Failure of the
timing analysis means that either faster logic gates must
be used or more parallel logic structures devised from the
equations.

9.3.2   Multiplexers

The implementation with multiplexers recognises the
fact that a multiplexer is a direct implementation of a K-
map. That is, each input represents one map square if the
state variables are connected to the select inputs of the
multiplexer. Therefore, having determined the entries for
the K-map as above, they become the functions that are connected to the multiplexer inputs. Fig 9.3 shows the connection diagram for the NC signal described in the K-map
of Fig 9.2. The advantage of a multiplexer approach is
that the design is quite flexible, permitting easier
changes once built and the hardware structure is simple.
The circuit can be fast since only a single chip delay is
involved in the feedback path. For VLSI designs a small
multiplexer can be configured from pass transistors and is
quite compact; it can offer an alternative to a full PLA
implementation. Multiplexers would not normally be used
for the output function since they are usually much simpler.

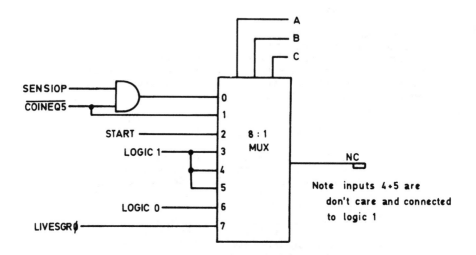

Fig. 9.3. Example of Multiplexer to implement NC

9.4   PROGRAMMABLE LOGIC ARRAYS

        This section covers the range of LSI components based
on the programmable logic array (PLA) structure, including
the field programmable logic array (FPLA's), the field
programmable logic sequencer (FPLS's) and the programmable
logic array (PLA's) used in a VLSI design. The first part
deals with the points common to all the types and then
each is discussed individually. Since the PLA has already
been described in some detail in chapter 8, only a summary
is given here.
        The PLA is a regular structure of gates providing a
set of programmable product terms which can be connected
to an array of OR gates forming sum terms. The output of
sum terms may have a programmable inverter. The PLA usually has many inputs and the number of product terms (PT's)
is considerable lower than all the possible combinations
of inputs: e.g. the type 82S100 PLA has 48 PT's and 16
inputs.
        The state table in Fig. 9.1b has 10 inputs but only
needs 16 product terms to describe it. Therefore the PLA
is ideally suited to the implementation of ASM designs.
The table is entered directly into the PLA. Lines in the
table with no active outputs (there are none in the example) are omitted. For programmable components there is

little virtue in trying to compress the table to reduce
terms, as suggested in chapter 7, for custom designs it may
be desirable to try to minimise the PLA area. From my
experience it is better to leave the table unminimised if
possible because the minimised solution will bear little
relationship to the original and if errors exist it can be
difficult to identify the offending term

Timing calculations for PLA implementations are easy
since there is only a single component involved, and they
are usually fast; e.g. the 82S100 is logically equivalent
to a 64K byte, 50ns PROM. PCB layout and circuit descriptions are simple but the PLA contents must be recorded. In
fact the pattern is almost useless; it is far more convenient to have the ASM chart. In view of the fact that custom-designed PLA's have already been covered in an earlier
chapter, the rest of this section describes various programmable PLA-type devices and their limitations and advantages. Further details are given in references (3) and
(4).

9.4.1   PLA's, FPLA's and FPLS's

For some years programmable logic arrays (PLA's),
which are masked programmed, and field programmable logic
arrays (FPLA's) which are electrically programmed, have
been available from a few manufacturers. A typical PLA has
96 product terms, 16 inputs and 8 outputs in a 28 pin package. A compatible FPLA contains half the number of product
terms. They provide a powerful component for use in complex hardware designs, being flexible in the case of the
FPLA, and thus making modification or correction of design
errors easy. The field programmable logic sequencer, FPLS,
includes a state register on the same chip, Fig. 9.4,
thereby increasing the logical power without requiring many
more pin connections. Therefore a 28 pin package with 14

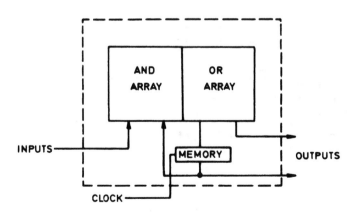

Fig. 9.4   Structure of a FPLS

inputs and 10 outputs (6 from the register, 4 direct from
the OR array) the AND array has a total of 20 input terms.
    Although quite expensive, probably due to their lim-
ited use by industry, these devices are cost effective in
low volume and produce a very flexible design. Two or
more state machines can be implemented in a single PLA,
especially if they share inputs. If two linked state machines are implemented in one PLA further economies can be
obtained by suitable merging of the two machines.

9.4.2   Programmable array logic (PAL)

    The PAL is an optimised variant of a PLA, which pro-
vides more inputs or outputs or functionality by reducing
the width of the OR array. Thus, whereas a PLA allows all
product terms to be connected to the OR array, the PAL
allows only a subset of all product terms to be connected
(see Fig. 9.5) by having a limited number of inputs on the
OR array terms. A wide range of configurations are available and a selection is given in Table 9.1. The table

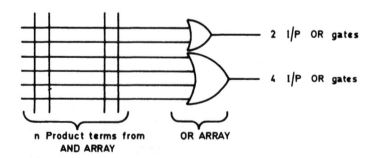

Fig. 9.5 Restricted arrangement of programmable
          array logic

INPUTS	OUTPUT	H/L	OR FANIN	REG	FEED BACK	3 STATE	TYPE NO PAL-
10	8	H	2	—	—		10H8*
12	6	H	4(2) 2(4)	—	—		12H6*
14	4	H	4	—	—		14H4*
16	2	H	8	—	—		16H2*
16	1	H/L	16	—	—		16C1
20	1	H/L	16	—	—		20C1
12	10	L	2	—	—		12L10
14	8	L	4(2) 2(6)	—	—		14L8
16	6	L	4(4) 2(2)	—	—		16L6
18	4	L	8(2) 4(2)	—	—		18L4
20	2	L	8	—	—		20L2
16	8	L	7	—	7	8	16L8
20	10	L	3	—	9	10	20L10
8,8(R)	8	L	8	8	8(R)	8(S)	16R8
10 6(R)	8	L	7(2) 8(R)	6	2(O) 6(R)	2 6(S)	16R6
12 4(R)	8	L	7(4) 8(R)	4	4(O) 4(R)	4 4(S)	16R4

Plus others with arithmetic functions capability.

* active low version as well

TABLE 9.1   PAL Types

shows the number of inputs available, some of them being
feedback signals from the outputs or the internal register
(R).   The output column shows the number of outputs available which may be from registers or gates.   The H/L column
shows the active level, low being the most common to permit
wired-OR connection.   The OR fan-in is shown, and where
mixed the number of gates is given in brackets, e.g. 4(2)
means two outputs are four-input OR gates.   The REG column
indicates if the chip contains registers.   The number of
feedback signals (i.e. internal connections back to the
AND array) is given indicating how many there are and
their source; (R) from register, (O) from an output gate.
The tri-state column defines whether the output has a high
impedance state.   For gate outputs the enable is derived
from the AND array direct while for the registers a single,
(S), external enable is provided.   The wide range of alternatives indicate how the design lacks generality and the
designer must match his requirements to the device selected.
Notice how all of them have a large number of inputs, this
being a key factor of array logic.

     To use a PAL it is necessary to ensure that the equa-
tions are arranged to match the PAL's available.   However
the extra functions available make them very powerful even
if used to implement only simple logical functions.

9.4.3 Expanding the component capability

     For large designs it may be necessary to expand the
capability of the single PLA or PAL component either by
using more of the same, or a mixture of PLA's and PAL's or
even using MSI and SSI components.   There are three ways
of expanding the component PLA or PAL; outputs, product
terms and inputs (see Fig. 9.6).

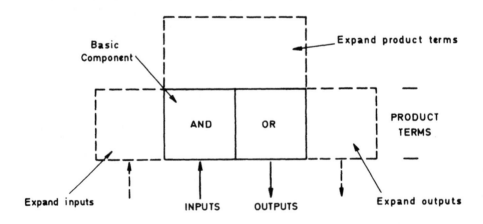

Fig. 9.6   Expanding a basic PLA or PAL component.

9.4.3.1 Outputs. PLA's and PAL's usually have a limited
number of outputs because providing more inputs is better
and the outputs can be expanded easily. Groups of exclusive output terms can be encoded inside the PLA and an external decoder used to generate each output term. This is
only beneficial when 6 or more outputs can be combined together. An alternative is to place another PLA in parallel
with the first connecting the inputs together, programming
both with identical product terms, as shown in Fig. 9.7a.
It is often better to rearrange the design to produce more
economical output utilisation: for example, state instructions can be implemented by external logic from the state
variable, avoid using conditional outputs and use state
assignments to make output functions simpler.

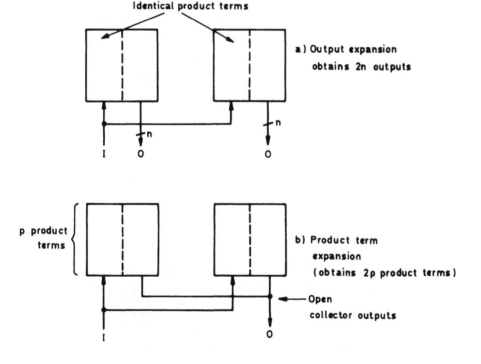

Fig. 9.7   PLA expansion methods

9.4.3.2   Product terms. Product term expansion is done to
increase the number of inputs to the OR gate array. With
the PLA this can be achieved by paralleling up two PLA's,
as shown in Fig. 9.7b. The PLA's must have open-collector
outputs providing a wired-OR function and the inverting
output function is used to give active low signals which
makes the open-collector connection a negative logic wired-
OR. For PAL's this method of expansion can be used to
increase the fan-in of the OR gate both within a single
component and using a second one, but active-low signals
are required in addition to open-collector outputs. In
some PAL's the output signals have tri-state controls which
can be programmed to a product term, but using this for PT
expansion can be difficult and should be examined carefully.

9.4.3.3   Inputs. The reason PLA's and PAL's try to provide a large number of inputs is because direct expansion
is impossible. To do this each AND gate in the AND array
has to have extra inputs, each being independently programmed. Thus to expand a 48 product term AND array requires 48 signals; obviously an impossible alternative.
However, input expansion is possible by implementation of
common functions or by use of exclusive inputs. The former
method requires the designer to recognise a multivariable
function which is used frequently enough to justify dedi-
cating an input to it and realising it outside as a separate product term, using gates, PLA or PAL if suitable. The
alternative is to multiplex inputs that are exclusive to
different ASM blocks, using the state variables as the
select signals. Obviously careful state assignment is necessary for this to be feasible.

9.5   ROM BASED MACHINES

        The increasing size and speed of PROM's and ROM's
makes it more attractive to use them as logic components
in a finite state machine. The ROM implements a truth
table directly with each output bit representing each entry
in the output side of the table. Therefore the state table
of the video game can be mapped into a ROM if enough inputs
and outputs are available. Direct mapping in this way is
only economical if the table has a limited number of inputs; for example the video game needs a 1K by 12 bit ROM
which is quite plausible with modern high-speed semiconductor memories. Even using todays technology the game
would use two ROM's to obtain the required number of outputs and the clock frequency would be restricted to around
500 nsecs. For designs with more inputs and a higher speed
requirement then a better approach than direct mapping is
required.
        To reduce the number of inputs to the memory it is
necessary to remove the more redundant inputs; i.e. the
qualifiers. Figures 9.8 and 9.9 show two alternatives.
In the first the address (state code) is stored in two link
fields which permits the selection of one of two next
states. This is performed by the qualifier which is itself

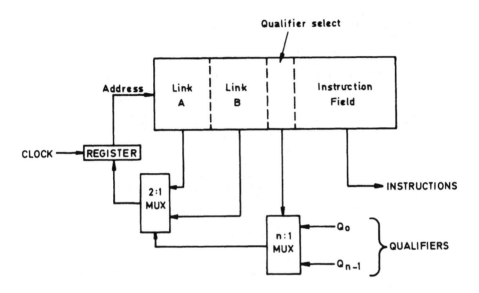

Fig. 9.8   Basic linkpath address ROM based machine

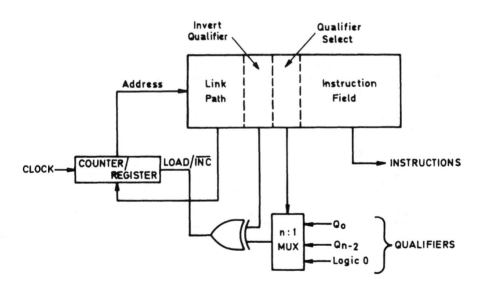

Fig. 9.9   Basic microprogram machine

selected by a ROM output field which chooses one to control the next state. The selected state is loaded into the
state register which provides the ROM address. The machine
therefore trades off width for locations; a very attractive
solution if many bits of address are saved.

The tradeoff does cost something. Each ASM block is
restricted to a maximum of two next states and a single
qualifier. The design may need serious modification to
accommodate this restriction. The result is that more
states will be needed, but they are not very costly. The
number of qualifiers is only limited by the size of the
multiplexer used. Thus the design is very compact and mapping the table to the ROM is relatively trivial. Fig.9.9
shows a variant on the same theme using a counter to give
one of the next state addresses. Here it is necessary to
ensure that one of the two states is one count more than
the current state. The second state is loaded from the
memory according to the state of the qualifier and the
inverter function. The logic 0 qualifier is to permit single next state transitions to be implemented and the inverter makes the assignment of state codes less critical.

Both methods can be enhanced by putting more external
logic in the feedback path to select more qualifiers at once
or to give greater functionality to the hardware. The system soon takes on the appearance of a microprogrammed
machine and then more conventional programming methods
should be adopted. The machine in Fig. 9.9 is perhaps the
primitive microprogrammed machine if the requirement is for
it to have a program counter (add 1 to the current address);
if not then Fig 9.8 qualifies as the simplest.

9.6   SUMMARY

It is informative to see that there is a continuous
spectrum of implementation choices from the simple combinatorial logic circuit to a fully-fledged microprogrammed
machine all being derived from the same design philosophy
based on an algorithmic description of the solution. It is
not difficult to see that programs running on computers are
an upward extension of the same theme. Here hardware implementations have been concentrated on and a wide range of
options have been illustrated. The arrival of semi-custom
and VLSI design methods has extended the options even further. As well as being a well organised method it provides
a standard approach that all designers can adopt thereby
making designs more understandable. The method provides a
good documentation base which also improves the portability
of the designs. ASM is not a replacement for the more
sophisticated methods of design being produced for VLSI
systems, but it does provide a workable method which may
suffice until these tools are available.

REFERENCES

1.   Clare, C. "Designing logic systems using state
         machines". McGraw Hill, 1973.

2.   Edwards, M.D. Private communication on paper to be
         published.

3.   Proudfoot, J.T. "The use of programmable logic
         arrays". Electronics and Power, 1980,
         26, 11, pp 883 - 887.

4.   Dagless, E L., Aspinall, D. "Feasibility study
         on the use of FPLA's". Final report
         to ACTP, contract no 3008, 1978.

Chapter 10

CAD and design automation

H.G. Adshead

10.0   INTRODUCTION

Today's circuits, chips and computers are so complex
that for most practical purposes it would be impossible to
create then without a considerable amount of use of
computers as a design aid.   In addition the concept of a
semi-custom approach to integrated circuits is one that
applies automated design techniques very effectively to a
specific class of chips.

This chapter discusses the reasons for and benefits of
using CAD.   Then, specifically in the context of regular
chips such as gate arrays and standard cell structures,
several classes of DA algorithms are described.

10.1   USE OF COMPUTERS IN DESIGN

There are 5 main ways in which computers can assist the
electronic design process (Fig.10.1).

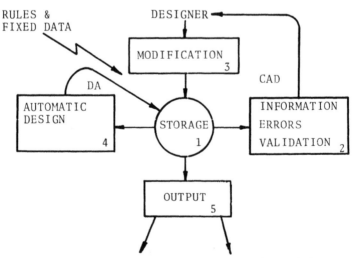

Fig.10.1   DA & CAD - The main ways in which computers
support design

Referring to the five numbered program areas in Fig.10.1, computers are used to store a model of the design (1) and can hence provide a very valuable design 'library' of the current and past states of the design.

Box (2) represents a whole class of programs which can provide the designer with useful views of the design either as listings or in graphical form. Also programs are used to analyse, evaluate or simulate the design and indicate possible design errors.

A good design system will provide a variety of techniques for editing or modifying the design (3). Design is an iterative decision-making process and ideally the data storage and processing mechanisms will rapidly follow up the effect of a small change to a large structure. The closing of the loop 2-3-1 (view-designer-modify-store) is CAD - Computer Aided Design - where essentially all design decisions are being made by the human designer. CAD can be very effective especially if the system can provide a response using devices such as CRT Graphical Displays.

Box (4) represents a further class of programs that actually contribute, expand or synthesise design data. This loop is technically very similar to the CAD loop but does not contain any human designer. These are the DA - Design Automation - algorithms. Sometimes these programs can be quite glamorous, eg. automatically laying-out an integrated circuit, fitting in 10,000 interconnect wires and creating a full set of production test data.

Box (5) indicates the set of programs concerned with outputting the final design data to the manufacturing and field areas. Most of this data for an IC will be in the form of NC - Numerical Control - tapes to drive the Mask Making and Test Equipment.

10.2 BENEFITS OF CAD

There are many benefits or motives for using CAD. They focus on labour saving (synthesis of design and managing the design process) and error removal (leading to faster design timescales). Perhaps the most significant benefit is that all output, documentation and NC data is consistent.

IC design is complex. Complexity is some function of scale or numbers of objects to be considered, the lack of regularity and the number of parameters of concern.

CAD and DA become essential in the IC design process in many ways. The most glamorous CAD tools relate to large analysis of complex parameters such as circuit and timing performance. The most effective DA tools apply to the automatic layout of large but regular and constrained circuits.

10.3 CAD FOR SEMICUSTOM IC's

This paper concentrates on the use of computers in designing semicustom IC's. In general this refers to a class of chips (say of 1000 to 4000 transistors) made up of small, standard, proven building blocks which may be safely interconnected by the user to create the desired custom circuit.

The most effective semicustom approaches are those where the user performs logical design using blocks from a standard library which have been carefully designed to be amenable to DA techniques, which usually means they are constrained, regular and structured.   DA algorithms are then employed to layout the full circuit automatically and produce all the associated test and mask generation data in a matter of a few hours.

Key semicustom technologies are gate arrays and cell arrays.   Gate arrays employ fixed transistors and power distribution and then a variable metal pattern is used to personalise groups of transistors into the logic blocks and add the interconnection pattern between these blocks.   Cell arrays employ standard circuit cells which are usually rectangular with one common dimension to allow a regular power distribution.   The DA algorithms layout these cells in a set of rows with a channel spacing just sufficient to allow all the necessary wire routing.

The remaining sections of this paper will briefly discuss the DA and CAD techniques used.

10.4   PLACEMENT

The placement problem is to assign a position to each block.   Associated problems are to allocate the chip I/O pins, cluster 2 or more sub-blocks into full blocks.   Sometimes the placement will be constrained with questions of timing and maximum allowed wire length.

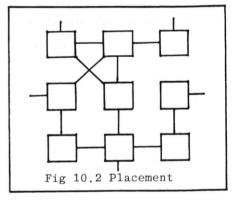

Fig 10.2 Placement

The best algorithms seek to minimise the total wire length while at the same time spreading out the wiring demand to minimise wiring congestion in any channel.   One of the best approaches to achieve this is selective partitioning into rows and columns where each sub-division is chosen on the basis of minimising the number of wire links crossing each cut line.

10.5   TRACKING or ROUTING

The routing problem is to assign a particular route or track to each desired interconnection.   Associated problems are to assign particular pins into or out of a block where there may be several logical or physical alternatives, to order a net or string of connections, to assign wires into channels, to allocate feed-throughs or vias in cases of limited transparency.   Sometimes the routing will be constrained so that no wire or net may exceed a certain percentage above the theoretical minimum wire length.

There are several algor-
ithms which have proved highly
successful.   In general a
good semicustom IC approach
will be matched to a particular algorithm.   There are 2
main classes of algorithm.

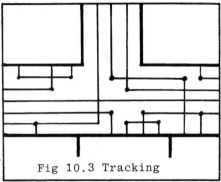

Fig 10.3 Tracking

The first involves
channel assignment where the
connections are first generally allocated into wiring
channels and then a separate
operation is carried out within each channel to assign the
actual wire routes.

The second class first sorts all nets or wires in some
order and then attempts to fit each in turn.   The best
techniques will guarantee to find a path if one exists -
such as line search or Lee's algorithm point search. Reference
(1) contains a very good summary of layout techniques together with many further references.

Cell arrays can usually be routed totally automatically
as can most gate arrays.   However for some dense circuits
gate arrays will produce some wiring failures.   Interactive
CAD techniques are needed to fit in these wires.   The best
systems will facilitate the temporary removal of wires or
part tracks while at all times retaining the full logical
integrity of the circuit.

10.6   TEST GENERATION

Test data may be created
as a CAD byproduct of simulation techniques using tools
such as testability analysis
and fault simulation.   Better
still, when the logic circuit
can be constrained somewhat,
a full set of test data may
be generated entirely automatically.   If all the state
elements (latches) are chained
into a serial shift register
with a regular clocking structure then the test generation
problem is reduced to the
remaining combinatorial
portion.

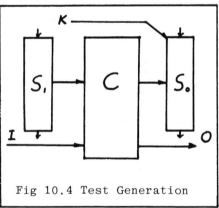

Fig 10.4 Test Generation

The D Algorithm postulates each possible fault condition in turn and then exhaustively explores and finds an
input pattern that will exercise the fault situation while
allowing evidence of the fault to propagate to at least
one output terminal or latch.   This pattern is then simulated and all other faults covered are ticked off.   The
process is repeated until all faults have been covered by
the set of test patterns.

10.7   INPUT & CAE WORKSTATIONS

Data capture and input to the DA system is a very important aspect.   There has been a considerable growth recently in the use of single-user graphical scientific work stations.   Logical design data may be captured by the use of an alphanumeric HDL (Hardware Description Language) or graphically by drawing blocks and logical interconnections on the screen.   The best systems are hierarchical and allow the creation and use of super blocks to describe a complete system in a well structured manner.   Full logical and rule checks are carried out.   For smaller circuits all design operations may be performed on the work station.   For bigger chips it is necessary to link the work station to a larger computer to carry out the more extensive calculations.

10.8   VERIFICATION & SIMULATION

For large circuits it is far better to check them out with simulation techniques than to attempt to build a hardware 'bread board'.   A good simulator will be multi-level - ie. will allow the simultaneous exercising of a circuit at different levels of detail.   The most appropriate levels for semicustom IC's are the gate level (AND,NOR etc.) and the functional level (M:= K+L etc.)   The simulator first constructs a very efficient simulation data model which is designed to propagate events as rapidly as possible.   The user provides a set of input events or stimuli and requests a set of nodes to be monitored.   The output response on these nodes is either displayed as the simulation proceeds or is stored on disc for later perusal in an interactive fashion.   For semicustom circuits it is not usually necessary to model time precisely.   A nominal time delay is often used for each block.

10.9   TIMING VERIFICATION

Simulation is not considered to be the best way of discovering timing errors as the worst-case path may never be exercised logically.   Timing verifiers and timing path checkers will cover all possible paths between clocked latches.   A good timing checker will take account of the precise loading delay of the actual track lengths used and the loading parameters of all points in each net.   Also timing assertions are made for all I/O pins.   Then for rising and falling edges the minimum and maximum delays (ie. 4 values) are calculated for all possible paths in one sweep through the circuit.   A second reverse sweep is then used to identify and diagnose the worst-case paths that could violate the clocking requirements.

10.10   CELLS & SPECIFICATIONS

Some DA systems use extensive CAD to design and verify the cell blocks.   Often this data is used to create the cell library and the specification documentation to be used by the semicustom chip designer.

10.11   CONCLUSION

The above should serve as a general introduction to the most used DA and CAD techniques in the context of semicustom design.   There are of course as many CAD tools as there are chip design methodologies.

Future semicustom and custom VLSI chips will employ a mix of gate arrays, standard cells, variable sized cells PLA's and regular structures.   All these require an increasingly sophisticated and fascinating set of integrated computer tools.

When all the design from high level to manufacturing data can be carried out within the DA box we will have achieved the true Silicon Compiler.

REFERENCES

1.   Soukup J., 1981, 'Circuit Layout'. <u>Proc.IEEE</u> Vol 69 No 10, Oct 1981, p.1281-1304.

Chapter 11

A review of simulation techniques

Keith Dimond

11.1 INTRODUCTION

The Dictionary definition of simulation is the "making of working replicas of machines for demonstration or for analysis of problems". In the context of digital design these models are created using mathematical models, within a computer, to investigate the operation of a digital system

(a) to check its logical operation

(b) to assess its performance in relation to its specification

(c) to investigate its operation under fault conditions

Simulation has been used for a long time in digital systems design. The capability to model hardware, which has yet to be constructed, has an obvious attraction to a design engineer. With the increased use of custom and semi-custom integrated circuits, the designer must be assured that they will work, before the large sums of money needed to fabricate such devices are hazarded.

11.2 LEVELS OF SIMULATION

For any digital system there are many levels of sophistication which might be adopted by a simulator. If we consider an analogy of a mountaineer. When the mountaineer starts his ascent, walking through the foothills he will see the great detail of the flora and fauna, but his awareness of the large scale geographical structures is limited. When he has scaled the peak, he will be unable to see the detail of plants and trees, but he will now be able to see the general form of the landscape over a wide area. It would be imprudent of the mountaineer to attempt the study of the small scale structure of leaves etc., from the top of the peak, and he would be handicapped if he only attempted to study the geographical structure from the foothills.

In exactly the same way the digital system designer must be aware of the differing levels of detail at which he may model his design and know the advantages and disadvantages of each different level. It is not surprising to consider that if more detail is required about the operation of a subsystem then this will require that extra detail be incorporated into the model. Similarly the extra detail will be bought at a cost, the cost of increased effort to take account of the added sophistication.

We will now look at a general categorisation of levels of simulation. As with any simple classification it is often the case that multiple entries will be required. This will become apparent when we review the simulators available.

At the highest level of simulation, there is the minimum of detail. The design may be considered to be partitioned into a number of sub-blocks, interconnected by some flexible data paths. In this black box description, it is possible to represent the action of each block by means of statements in a high-level language. The inter-communication between blocks is achieved by passing complex data structures. This level of modelling allows the overall organisation to be investigated, and as a result of such modelling modifications to the partitioning may be carried out. In addition it is possible to predict where problems may occur in terms of data flow. We will call this level of modelling the systems level.

The next lower level of simulation is where each of the sub-blocks is decomposed into some functional structure. Typically this will consist of data highways and large scale logic blocks such as arithmetic logic units, registers etc. We will call this level the register-transfer or functional level, and it corresponds to the machine code or microcode of a processor. In such a simulator the information which flows around the system is binary vectors, in a suitably encoded form and this is processed by blocks of combinational logic. In addition to these highways and processing blocks a series of control signals have to be modelled which determine the precise sequence of transfer.

The next lower level of simulation is where the registers and combinational blocks have been expanded into their gate equivalents. This type of simulator is typically called a Logic Simulator. A wide variety of simulators of this type are available, varying greatly in their sophistication and facilities offered. The simplest is only capable of simulating idealised gates with simple fixed propagation delays; in contrast, the most sophisticated would have a complex representation of nodal values and the ability to vary propagation delay of individual gates according to their loading.

The lowest level of simulation which we will consider is that of the circuit simulator. Here the system variables of the simulation are voltages and currents. The action of the simulator is now to model the changes in these parameters in response to changes in the excitations. Since these simulators solve the circuit equations, it is possible to obtain precise details of the rise and fall characteristics of the gate outputs as functions of their loading.

Before proceeding further it would be as well to make some observations. The first is that we have seen how as we descend in level of simulation the detail increases. This will mean that the effort required to model the behaviour of one node of the circuit will increase. This will therefore mean that the cost of the design phase will increase proportionately. It should be realised that cost is not the sole consideration, for with a modest sized system and limited computing resources it may not be possible to simulate the system. This might happen because the memory available on the machine is not sufficiently large to be able to store all the data associated with the model. Alternatively it might be possible to simulate the system, but due to requirements of other projects it may not be possible to produce complete results in a reasonable time scale.

For a fixed quantity of computing resources, at the highest level it will be possible to model a system with limited detail for a long period of time. At a lower level it may only be possible to model the complete system for a short period of time or to model a part of the design for the required period of time.

There is a further point which must be considered. At low level simulation very much more data has to be provided so that modelling can proceed with the required precision. There is a difficulty in that not all the parameters for the device may be known, or they may not be known to sufficient accuracy. We now find a situation where large amounts of computing resources may be used generating very accurate results, which do not quite relate to the device or sub-system under consideration.

11.3   A REVIEW OF TECHNIQUES EMPLOYED IN LOGIC SIMULATION

Models of digital systems may be constructed either in hardware or software or in a mixture of both. Hardware models can be useful, but the effort required to construct and modify such a model can be quite significant. It is currently the practice of some semiconductor houses who offer gate arrays to require a TTL mock-up as well as the description of the design. Most of the simulators available today are software based, this means that the specification of the digital system has to be coded into a suitable form for input to the simulator system. The precise form of input varies from simulator to simulator, in some it may be a numerical and alphabetical coding of the gates or modules and their interconnection, with others much more structure may be available.

This input specification is then processed and fed to the simulator to produce the software model of the system. There are two main methods of operation used by simulators, these are normally referred to as Compiled-Code, and Table-Driven. Once the software model is available it may be exercised with appropriate input stimuli and the resulting outputs produced. The designer will typically exercise the model with a number of different input sequences to verify the design.

11.3.1   Compiled-Code Simulators

In this method of operation, the specification of the design is translated into machine operations of the processor on which the simulation is to be performed. For simple gates the logical operations of the native machine code are perhaps sufficient, for other more complex modules then subroutines are required to represent their operation. In addition to generating the code the translation systems must ensure that the order of evaluation of sub programs etc. is correct. The translator therefore determines the position of the various modules in terms of level in the circuit not the position in the source description. In this way the model of the system may be updated by a single pass through the code. When dealing with sequential circuits it is normal to iterate several times through the code until a stable condition is reached.

This approach is simple but makes it rather difficult to model accurately the operation of large circuits. Thus a different approach is normally used.

11.3.2 Table-Driven Simulation

With this type of simulator, the description of the circuit under consideration is translated into a set of data structures. These structures represent the connections and state of the system being simulated. The action of simulation is then performed by a program which accesses the appropriate elements of the data structure and produces new system values which are then written back into the appropriate structure. This method of operation is sometimes called interpretive simulation because of its similarity to interpretive execution of a high-level language.

This method of simulation has considerable advantages in that complex models and interaction can be dealt with, and gates can be modelled in a much more realistic way, ie non-uniform rise and fall times.

11.3.2.1 Timing model. Time is quantised in a simulation of a digital system. Signals within the system may only change their values at these time intervals. The accuracy of simulation is very closely related to the model which is adopted for time. Simulators exist with a variety of different models.

The simplest method is that of zero delay, which implies an assumption that the propagation delay through the gate is zero. This means that the simulator simply evaluates the logic equations of the system. No information is available on time and hence this form of simulator cannot detect malfunctions in the time domain.

The unit delay model is more realistic in that it assumes that all gates have a non-zero propagation delay time, but this time is assumed to be equal for all devices. This extension does enable time delay malfunctions to be detected but with a simple model the results of the simulation have to be interpreted carefully. Such simulators need to have a mechanism for ordering the modelling of individual devices.

If an attempt is made to model time more accurately then a number of different factors have to be taken into account. The propagation delay of a gate will be affected by gate loading and wiring capacitance. In addition it will be necessary to take account of the transient response of the gate ie rise and fall times, these will be different and will again be a function of gate loading. There is a further complication as one tries to refine the time model. With a real gate, there is a minimum pulse width below which the output of the gate will not respond. An accurate model of the gate must be able to represent this situation. It will be appreciated now that with a well-developed time model, scheduling plays an important part. The mechanism which is normally used is that of a time queue. For each change of input, it is possible to calculate the time at which the gate output will change. Data describing this future transition will be put into the time queue at the appropriate point and thus when the simulation time has reached this value, then the change in output will be actioned. For a long simulation, this would imply a time queue which had a prohibitive length, so it is normal to use a cyclic time queue, sometimes called a time wheel, so that slots in this queue could be re-utilised.

11.3.2.2 Signal models. It has been shown in the previous section how the time model for a simulator can be developed. The same development is possible for the signal model, thus in the simplest possible case two values are needed to model the state of a node in a circuit. The main disadvantage is that of initialisation, where each signal node must have an initial value and if all values are zeroed, then there will be logical inconsistencies. Initialisation could be performed by hand but this again is prohibitive for a large circuit. Furthermore if there are spikes or other transient malfunctions then it is not possible to represent these situations.

An advance on this is to use a three value signal model, with the introduction of an unknown value 'X'. This allows the initialisation problem to be dealt with and enables the simulation to represent in a reasonable way uncertainties due to races and hazards.

The next level of sophistication is to adopt a five value model, this consists of the two true binary states $(0,1)$, a rising and falling transition and the undefined condition. This latter value is used at initialisation and whenever the result of a operation yields an undefined or hazard condition.

11.3.2.3 Operation of simulator. The description of the system to be modelled is translated and stored in a number of data tables. There would be a circuit description table which will contain information on the function of the gate. Another table will contain the current value of each node in the system. For each of the gate types in the simulation, there would be a table in which is stored details of the number of inputs and outputs, and the values for rise, fall and delay times. Finally there would be a data structure into which had been input the excitation for the system.

The simulation would proceed by setting the primary inputs of the system to their initial values and from these inputs the transitions which are subsequently caused by these changes in inputs are put into the time queue, and time advances.

11.3.3 Hybrid Simulation

It will be appreciated now that a large amount of computing has to be performed to model the response of a digital system to just one change in input. It is therefore not surprising that simulation of large scale digital integrated circuits for an appreciable number of input transitions can take many hours to complete. In some areas this time factor is limiting the completion rate of the design and the computing resources employed represent a large cost overhead on the design process. Over the last few years there has been an increasing interest in hybrid simulation as a way of reducing the time required to simulate large circuits.

In this approach the simulation of the gate actions etc. is performed by special purpose hardware. This hardware has details of the interconnections and logic functions to be performed stored in memory. Similarly the values of the signals at each node are also stored in memory. In simulating a system, the hardware executes a series of commands, each command normally being equivalent to one gate writing the results to memory. There is also a host processor which is used to process the specification of the system and load this translated form into the memory of the hardware unit. In addition it extracts the signal outputs which are to form the results of the simulation. The basic operation of the logic simulation hardware can be explained as follows. Each gate is represented by an instruction, which specifies the address in memory of the operands and the function to be evaluated. Thus these operands are accessed from say Value Memory A and then the results of the operation is written to Value Memory B. When the next cycle of operation of this gate is modelled, then this time the operands are accessed from memory B and the results are set back into memory A.

In order to improve performance it is normal to introduce parallelism, thus it would be normal to have a number of these logic simulator hardware units generating outputs of gates in parallel. This introduces a complication since it is necessary to arrange for the output of a gate simulated in one processor to be needed as an input in another processor. To achieve this a multiway switch is necessary.

In addition to translating the source language into a suitable form for the hardware processor, the host processor will also perform allocation of gates to processors to ease the problem of scheduling.

11.3.4 Circuit Simulation

When implementing digital systems using standard families of logic functions (eg TTL, 4000 series CMOS and so on) there is seldom need to use simulation facilities other than those already described. One of the advantages of using semi-custom design techniques is that if the manufacturer has paid attention to the design of his cells there will be no need to investigate their internal operation and hence conventional logic simulators may be employed.

No review of simulation would be complete without reference to circuit simulation. So far as digital integrated circuits are concerned circuit simulation is needed to analyse the detailed internal behaviour of a gate or a collection of gates. Circuit simulators represent the system by means of their electrical network equations. In solving these equations it is possible to determine the currents flowing in branches and the voltages between nodes. In addition to passive components circuit simulators must also contain models of active devices. For a model of a transistor to be useful in integrated circuit simulation, it must be possible to incorporate into the model information about the dimensions of individual devices. It will be appreciated that an accurate model of a transistor will require large quantities of data concerning resistivities and capacitances associated with the various layers used in its physical construction.

It has already been mentioned that detailed simulation of a circuit requires massive quantities of computing resources. This means that in practice it is only possible to model part of a design at the circuit level. The problem then is how to define valid input sequences to this circuit to enable realistic operation to be simulated. One way of dealing with this situation is to employ a so-called 'mixed-mode-simulator'. In this it is possible to combine together various levels of simulation in one model. Thus the part of the design which has to be investigated in detail will be represented at the circuit level, while the remainder of the design (which supplies the inputs and uses the outputs of this part, and whose operation does not need to be modelled in great detail) are modelled by a conventional logic simulator. In this way it is possible to obtain the best of both worlds. It is likely that simulators of this type will be developed extensively in future.

11.4   A SELECTION OF SIMULATORS AVAILABLE

It is not possible in a short space to explain in detail how a number of simulators may be used. This section will list some simulators currently available in the U.K. and then give an example of one logic simulator and one circuit simulator.

HARTRAN[1]: Developed by GEC Hirst Research Laboratory. This is a functional simulator, capable of simulating clocked synchronous systems. The structure of the language is very similar to FORTRAN; it can be mounted on quite small machines.

HILO[2]   : Developed by Cirrus Computers Limited and Brunel University. This simulator supports most logic primitives and has a number of interconnection types. Available on a number of different machines.

DIGSIM[3] : Developed by SAAB-SCANIA and Linkopping University. This simulator is stochastic in operation giving a most likely time for signal transition and also a range of times over which transition may occur.

SPICE[4]   : Developed at the University of California Berkeley. This is a widely-used circuit simulator available on a variety of machines and depending on machine size capable of dealing with different numbers of nodes.

DIANA[5]   : Developed at the Katholieke Universiteit Leuven, Belgium. This is a mixed-mode simulator capable of dealing with logic, timing and circuit simulation.

11.4.1 Examples

The first example is a very simple one, of a D-type flip-flop. The logic configuration is given in Fig. 11.1. The specification of the circuit as a HILO program is shown in Fig. 11.2. This model is exercised by a set of waveforms described in Fig. 11.3, and the output of the simulation is shown in Fig. 11.4.

The second example is that of a circuit simulation of a simple two input NAND gate. The schematic for this is shown in Fig. 11.5 and the coding in SPICE is shown in Fig. 11.6. In SPICE signals and connections are referred to by node numbers. Hence it will be seen in line 3 that V_{DD} is connected across nodes 6 and 0 and has a value of 10 volts. The pull-up transistor is designated MU1 and is specified accordingly and the series transistors are MD1, MD2.

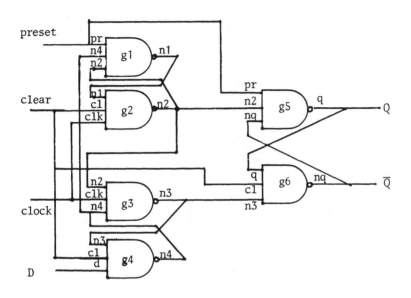

Fig. 11.1   Logic configuration of D-type flip-flop

```
 1 : cct sn7474 (q,nq,d,clk,cl,pr)
 2 : ** Asynchronous D type flip flop with preset and clear.
 3 : **
 4 : nand (22,15)
 5 :       g1 (n1,pr,n4,n2)
 6 :       g2 (n2,n1,cl,clk)
 7 :       g3 (n3,n2,clk,n4)
 8 :       g4 (n4,n3,cl,d)
 9 :       g5 (q,pr,n2,nq)
10 :       g6 (nq,q,cl,n3);
11 : unid n1 n2 n3 n4 q nq;
12 : input d clk cl pr.
```

Fig. 11.2   HILO specification of D-type flip-flop

```
1 : waveform sn7474 tst
2 : stimulus d,clk = 0
3 :        pr,cl = 1;
4 : 0 clk = change0(500,1000,1500,2000,2500);
5 : 0 pr = change1(250,350);
6 : 0 cl = change1(100,200,1800,1900,2700,2800);
7 : 1250 d = 1;
8 : 3000 finish.
```

Fig. 11.3   Excitation waveforms

```
              C
              L P C         N
              K R L D     Q Q
    TIME
           0  0 1 1 0     X X

         100  0 1 0 0     X X
         122  0 1 0 0     X 1
         137  0 1 0 0     0 1

         200  0 1 1 0     0 1

         250  0 0 1 1     0 1
         272  0 0 1 0     1 1
         287  0 0 1 0     1 0

         350  0 1 1 0     1 0

         500  1 1 1 0     1 0
         537  1 1 1 0     1 1
         552  1 1 1 0     0 1

        1000  0 1 1 0     0 1

        1250  0 1 1 1     0 1

        1500  1 1 1 1     0 1
        1537  1 1 1 1     1 1
        1552  1 1 1 1     1 0

        1800  1 1 0 1     1 0
        1822  1 1 0 1     1 1
        1837  1 1 0 1     0 1

        1900  1 1 1 1     0 1

        2000  0 1 1 1     0 1

        2500  1 1 1 1     0 1
        2537  1 1 1 1     1 1
        2552  1 1 1 1     1 0

        2700  1 1 0 1     1 0
        2722  1 1 0 1     1 1
        2737  1 1 0 1     0 1

        2800  1 1 1 1     0 1
```

FINISH TIME = 3000

Fig. 11.4   Output of simulation

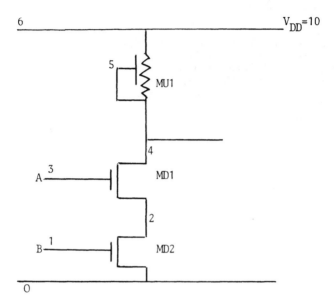

Fig. 11.5   Schematic of NAND gate

```
BASIC NAND GATE (.MODEL VALUES FROM SUMC)
.WIDTH IN=72 OUT=120
VDD 6 0 DC 10
.DC VS1 0 10.0 0.5
MU1 6 4 4 0 DEP(L=16E-6 W=4E-6 AS=48E-12 AD=24E-12 PD=24E-6
+PS=28E-6 NRD=0.75 NRS=0.75)
MD1 2 1 0 0 ENH(L=4E-6 W=8E-6 AD=48E-12 AS=48E-12 PD=28E-6
+PS=28E-6 NRD=0.75 NRS=0.75)
MD2 4 3 2 0 ENH(L=4E-6 W=8E-6 AD=48E-12 AS=48E-12 PD=28E-6
+PS=28E-6 NRD=0.75 NRS=0.75)
.MODEL ENH NMOS(LEVEL=2
+VTO=1
+KP=2.4E-5
+GAMMA=0.8
+PHI=0.57
+LAMBDA=2E-2
+CGSO=2.6E-10
+CBD=1.0E-15
+CGDO=2.6E-10
+CBS=1.0E-15
+TOX=1.0E-7
+PB=0.8
+XJ=0
+LD=0
+UCRIT=1E7
+UEXP=0.14
+UTRA=0
+NEFF=1)
.MODEL DEP NMOS(LEVEL=2
+VTO=-3.5
+KP=2.4E-5
+GAMMA=0.8
+PHI=0.57
+LAMBDA=2E-2
+CGSO=2.6E-10
+CGDO=2.6E-10
+CBD=1.0E-15
+CBS=1.0E-15
+TOX=1E-7
+PB=0.8
+XJ=0
+LD=0
+UCRIT=1E7
+UEXP=0.14
+UTRA=0
+NEFF=1)
VS1 3 0 PULSE(0 10 10NS 35NS 30NS 50NS 200NS)
VS2 1 0 DC 10
.TRAN 5NS 300NS
.PLOT TRAN V(4) V(3)
.PLOT DC V(4) V(3)
.END
```

Fig. 11.6 Spice model of NAND gate

11.5   CONCLUSIONS

It has not been possible in the space available to give a detailed account of the whole topic of simulation. The aim instead has been to illustrate the various levels at which simulation can be carried out and to highlight the major areas of interest to the designers of semi-custom integrated circuits. Simulators which are currently available provide the required results, but there is still a need to develop better user interfaces. By this is meant the format of specification of the design for input, and the way in which results are displayed. Of particular importance with regard to the input specification is its compatability with previous stages in the design process, since the aim must now be not simply to produce a design tool which operates in isolation but which is part of an integrated design system. This design system would take as its input a high-level specification of the system and then all the remaining stages take place with the minimum of designer interaction.

Acknowledgements

The author would like to acknowledge the support of SERC for the provision of simulation facilities and Mr. W.A.J. Waller for assistance with examples.

References

(1) Hartran, A Digital System Description Language.
G.E.C. Hirst Research Centre, Wembley, Middlesex.

(2) HILO-2 Users Manual.
Cirrus Computers Ltd., 29/30 High Street, Fareham, Hants.

(3) Digsim Users Manual.
Department of Electrical Engineering, Linkopping University, Linkopping, Sweden.

(4) Users Guide for Spice.
University of California, College of Engineering Department of Electrical Engineering and Computer Sciences.

(5) Diana Users Manual.
Department of Electrical Engineering, Katholieke Universiteit Leuven, Belgium.

Chapter 12

Partitioning, placement and automated layout

D.J. Kinniment

12.1   INTRODUCTION

An essential part of any Design Automation System for
semi-custom i.c. design is the autolayout software whose task
it is to translate the structural information provided by
the designer, usually in terms of a textual representation
of a logic diagram, into mask descriptions.

For both economic and technical reasons it is desirable
to reduce the number of chips used in a system to a minimum
with the eventual aim of implementing the whole system on
one VLSI chip.   In general, however, this is not possible
and the designer is left with the necessity of partitioning
the system over a number of chips.   After this task has been
performed, each subsystem must again be decomposed into an
interconnected network of smaller units, and so on until a
level of primitives is reached which will be mapped directly
onto the silicon.

The partitioning process described above is carried out
manually but is usually followed by an automated or semi-
automated layout process consisting of placing the primitives
on the chip in a set of locations whose positions are chosen
to ease the problem of routing the connections.   Since the
ease of routing is very dependent on the placement, some
autolayout programs deal with both together, however re-
evaluation of placement when the routing proves not to be
optimum can be costly in computation time.   For this reason
the two processes are often separated with manual intervention being allowed at the routing stage if a placement is
obviously inappropriate.

12.2   PARTITIONING

There is no clear cut way of measuring the effectiveness
of a given partition of any system, and presently no algorithm exists which can be used to decompose a design into the
optimum set of subunits, but there are a number of objectives
to the process, against which a particular partition can be
assessed.

Some of the more important of these are described below.

12.2.1 Regularity

It is obvious that economies can be made if a design can
be implemented by connecting together a number of identical
chips rather than the same number of different chips, but
this is simply an illustration of the fact that the total
complexity of the problem has been reduced by a partition
which reduces the number of different kinds of subunits. The
total design time will be lower if there are fewer subunits
and economies which are not necessarily related to the wafer
fabrication costs can be made which also apply to partitioning within the chip as well.

A good measure of this reduction in complexity is the
regularity of the design; the ratio of total design area to
individually crafted parts of the design.

12.2.2 Number of Interconnections

The number of connections between subunits in a partition is also related to its complexity, and hence design
difficulties also increase because of this in an inappropriate partition. More importantly the cost and performance
of an integrated circuit are adversely affected if the number
of connections it requires to the outside world is large,
and similarly partitions internal to the chip which result
in a large number of interconnecting wires, will involve increasing penalties in cost and performance of the chip as
the technology progresses.

This is because the reduction of dimensions achievable
by improved photolithography is likely to be used by
increasing the number of components on each chip.

The effect of 'stuffing' as many components as possible
on a chip will have important implications in design, and
here it is necessary to consider the interconnection path
between two gates. If the conductivity of the connection
material is high, the connection acts as a transmission line
and its delay is proportional to its length. Thus if the
length is scaled down the delay is also scaled down, since
capacitance is proportional to dimensions. Two problems
prevent this ideal state of affairs:-

(a) As the dimensions are reduced resistance of a
     connection increases because cross-sectional
     area reduces faster than length.

(b) The average connection length on a 'stuffed'
     circuit is similar to that on an unscaled chip
     of the same size because many connections
     still have to cross the entire chip.

These effects lead to a delay which is often dominated
by the resistive and capacitance time constant R.C., giving
a time delay which increases with scaling since R increases
and C remains constant. Figure 12.1 gives one comparison
between a 10mm length metal connection delay and a gate
delay for a range of feature sizes from $\lambda=5\mu$ to $\lambda=0.1\mu$. It

can be seen that above $\lambda=1\mu$ the system performance is dominated by the gate delay but below that figure it is the connection that makes the most important contribution.   A polysilicon connection would of course present far worse
problems.

10mm connection - metal

λ	5μ	2.5μ	0.5μ	0.1μ
R	25Ω	100Ω	2.5K	62.5K
C	4pF	4pF	4pF	4pF
RC	100pS	400pS	10nS	250nS
Gate	5nS	2.5nS	0.5nS	0.1nS
		←	→	

Performance dominated	Performance dominated
by circuits	by connections

(Polysilicon connections have resistance approximately
100 x R)

Fig.12.1   Interconnection performance

The importance of the connections rather than the gates
at VLSI complexities can also be seen in measurements made by
Donath et al (1) who showed that the total wire length in a
logic chip tends to be a function of the level of integration
as indicated in Fig.12.2.   In this figure average line
length of connections increases approximately as $(circuits)^{\frac{1}{2}}$
so that the area occupied by all the connections increases
as $(circuits)^{1\frac{1}{2}}$ and a point is reached where the majority of
the silicon area is occupied by connections rather than
active devices.   It is clear then that at VLSI levels design principles will have to change away from optimisation
of the use of components towards effective management of the
interconnections.
    This work is based on an observed partitioning relation-
ship known as Rent's Rule, whereby the average terminal
count T, is related to the block count C and two constants A
and p as:

$$T = A.C^P$$

Taking this as a starting point Donath and Mikhail (2) have
calculated the size of wiring channels required to interconnect a series of gate arrays of different sizes and
aspect ratios, and compared the results to measurements in a
study at IBM.   These results show that calculations based
on Rent's Rule provide a useful guide to the wiring channels
needed in a ULA.

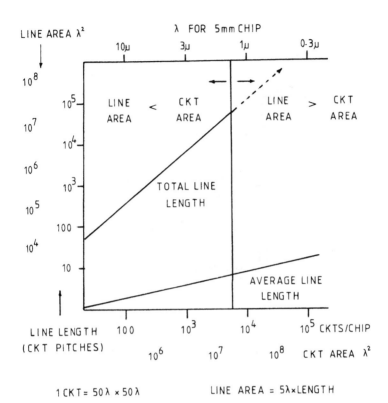

Fig.12.2   Relative connection area

λ = minimum feature size

12.3   AUTO LAYOUT

In the design of a ULA partitioning of the system proceeds first by identifying the subsystems which will be located on each chip, and then by progressive refinement of each subsystem until gate-level primitives are obtained which can be directly realised.   The place of autolayout in a typical ULA design system is shown in Fig.12.3.   Here the layout proceeds from a gate-level description to the place- ment of modules and their respective interconnections via simulation to verify the correctness of the decomposition of the design.

The IC designer encodes a logic diagram using a collection of available gates and macros which are either user defined or exist in an internal library.   The gate- level description in its simplest form consists of a series of component types with corresponding I/O signal names having the general format:

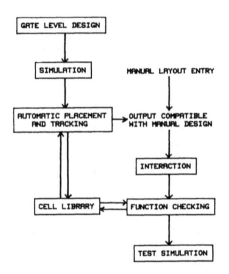

Fig.12.3   Overall Design System structure

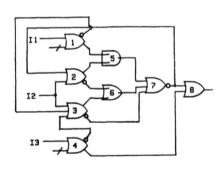

1	ORN	N1,O1	1	I1
2	ORN	O2,N2	2	N5,I2
3	ORN	O3,N3	3	I2,N1,N4
4	ORN	N4,O4	1	I3
5	WAND	W5	2	O1,O2
6	WOR	W6	2	N2,O3
7	NOR	N7	3	W5,W6,N3
8	OR	O8	2	N7,O4

Fig.12.4   A simple network and its encoding

Id No.   {position} Comp. type Output(s)

   No. of inputs Input(s)

   where {...} denotes an optional

   parameter.

Fig.12.4 shows part of a simple network and its encoding for
input to a design system for an ECL gate array.
   In order to place the component types it is necessary
to determine the area each component occupies by performing
an initial comparison with the gate library.

12.3.1   Placement

   Placement software in the system described above positions the gates and macros according to their strength of
connectivity to other gates.   The designer may also have
the facility of manually prepositioning gates by defining a
gate's position in the input specification.   This allows
the user to interfere with the placement process and observe
the effect on other gates and the subsequent layout of
interconnections.   An internal representation of the logic
diagram is used in conjunction with the library to determine
the interconnection net list.   The library is then used to
retrieve information defining the function of the gate and
co-ordinate information defining the end points of the nets.

12.3.1.1   Placement methods.   The main aim of the placement
algorithm in ULA and other semi-custom structures is to
provide the basis for the following routing process. Usually,
the wiring channel space allocated between rows and columns
of gates is a compromise between economy in the use of
silicon and the space needed to give a high probability of
successful layout as given by (2).   Because any electrically
valid design can be entered it may not be possible to route
all the connections even if the placement is good, and in
general it is difficult to estimate simply how good a given
placement is until the actual routing has been done.
   The computation required to investigate all possible
placements is prohibitive because the number of different
placements is proportional to n! where n is the number of
modules to be placed in an n-gate ULA.   Most methods
attempt to estimate the 'goodness' of partial placements as
the process progresses, choosing only the 'best' module to
place next in the 'best' location.   This involves continually estimating the 'cost' of selecting a module and
placing it in a particular location.   The process of cost
estimation must be fast in order to explore as many placement variations as possible, but must also bear as close a
relation to the actual routing process as possible to be
realistic.
   One method is to represent the cost of a partial placement by the total wire length, because this tends to minimise
wire length on the chip.

Unfortunately this also tends to bunch components in a relatively small area, leading to difficulties of congestion in the wiring channels in that area, as shown in Fig.12.5.

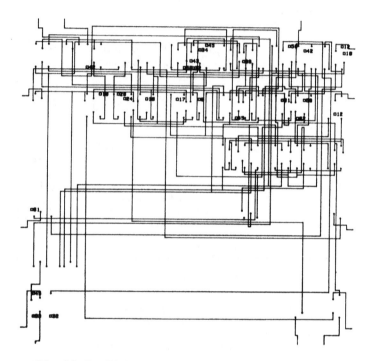

Fig.12.5   Placement according to wire length

It is clearly better to attempt to minimise the congestion in the wiring channels rather than the wire length itself, and many systems now do this, using weighting functions which depend on the probability of a wire using a particular channel and the likely occupancy of that channel by other wires.

Whilst these methods produce adequate results they are still inferior to the human designer in some situations, particularly where there is an easily identifiable structure to the design as shown in Fig.12.6.

Here a simple 12 x 12 array of gates is to be laid out on a 144 cell ULA, but since each gate is equally interconnected to every other gate on its row or column it is not possible for the algorithm to discriminate between many of the unconnected gates, and differences in placement costs are also small once a module has been selected.

Clearly the structure inherent in the design has been ignored by the placement algorithm and the penalty for this can be clearly seen in Figs.12.7 and 12.8 which show the

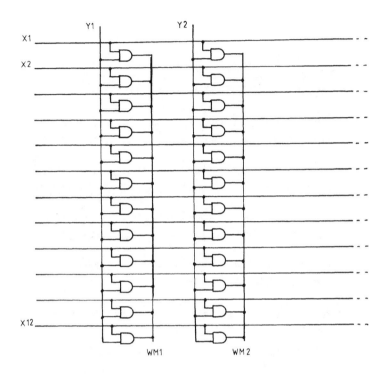

Fig.12.6   Regular test array

results of routing the autoplaced version of this problem
(Fig.12.7) and a hand placed version which takes account of
the two dimensional structure.

One   approach to improvement of   placement is the 'brute
force' method of increasing the computational resources to
explore more placement possibilities more accurately, but
this is a method not usually available to companies other
than IBM.

Most design automation systems therefore combine the
accuracy and analysis of a relatively simple algorithm with
interaction by the designer to provide the insights necessary
to explore any structure in the design.

12.3.2   Routing

After placement the routing software determines the
precise paths of the interconnections between points. Again
this is a problem in which the very large number of connections required interact with each other in a way which makes
the exploration of all possible paths impractical for a ULA
of present-day complexity.

At the simplest level there are many algorithms and

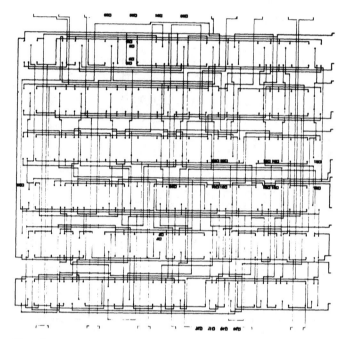

Fig.12.7   Auto placed array, minimum congestion
            algorithm

heuristics which have been used to join a net of points.   A
description of all these methods and their relative methods
is beyond the scope of this chapter, and only the essential
points of a few will be outlined.

12.3.2.1 The Lee path connection algorithm.   Perhaps the
most commonly used method for finding the shortest path
between two points was described by Lee(3).
        In essence this consists of choosing one point as the
source, S and the other as the target, T.   The area to be
explored to find the path is divided into a grid of cells
each of unit size, and the source cell marked with an
integer.   This marked cell is then placed into a list
called the frontier cell list and all unmarked neighbours of
the frontier cells are then marked with the next integer.
When no more unmarked neighbours exist, the old frontier
cell list is deleted and replaced by the list of newly
marked cells.   The process is then repeated as shown in Fig.
12.9 until the target is hit.
        If the frontier cell list has no members at any point
before the target is reached, there can be no path between
the source and target.   Lees algorithm thus guarantees that
if a path exists it will be found, and that the path found

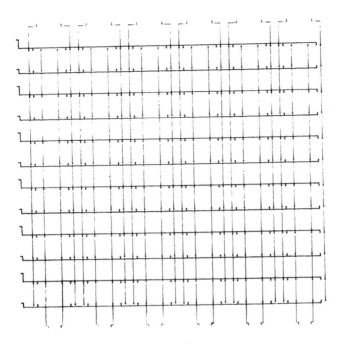

Fig.12.8 Hand placed array

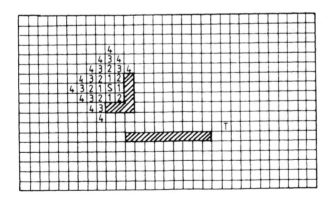

Fig.12.9 Lees algorithm in two dimensions

will be the shortest since at any time the minimum path
length to every frontier cell is equal.
   The actual path can easily be retraced by following
back from the target to the source looking at each point for
the previous integer in the sequence.
   The reason for the ubiquity of Lees algorithm is its
adaptability. It can easily cope with multipoint nets by
expanding from all points simultaneously and retracing in
both directions from a contact point, and can also be
modified to expand faster in one direction. This latter
modification facilitates X-Y routing on two-layer ULA's.

12.3.2.2 Heuristics.   Cellular routers such as Lees
algorithm may explore very large areas of board before a
successful connection is found and since the area is
explored on a cell by cell basis the computation required
will be considerable.   If the number of connections is
proportional to the number of gates n, and the average
length proportional to $n^{\frac{1}{2}}$, the area explored per connection
will be at least proportional to n and the total layout time
at least proportional to n^2.
   This may be worse if the connections are difficult and
most of the i.c. area has to be explored to find the
connections.
   Usually, however, about 75% of connections are
relatively simple involving nothing more than a straight
connection, or one with at most one or two bends. A heur-
istic approach which attempts to find these simple connections quickly by extending parallel lines from the two points
and connecting between the lines with an orthogonal line was
first described by Aramaki (4).   This has the advantage of
following an x-y path which as well as being quick to
compute is less likely to block subsequent nets than the
complex paths sometimes produced by Lees algorithm.

12.3.2.3 Channel routers.   Rather than looking at each net
in turn and dividing the area available into a grid of cells,
many recent routers examine each row or column in order to
pack connections optimally in the wiring channel formed between the rows or columns.   An early method was described
by Hashimoto and Stevens (5).   Here all the nets partici-
pating in the use of a particular channel are considered at
the same time, and the aim is to minimise the channel width
required to provide the connections and through routes.
   A simple way of doing this would be to assign those
connection segments traversing the greatest channel length
to a position near the centre line whilst at the same time
attempting to fit other short segments into each position to
utilise the smallest possible number of positions across the
channel width.

12.3.3 Hierarchical routing

   In a completely global router which tackles the whole

routing problem in one pass using a net routing algorithm
such as Lees algorithm, the amount of computation required
is at least proportional to the square of the number of gates
in the ULA.   Considerable savings in time can be made by
the use of local, channel routers which consider the
detailed wiring of a relatively small area in a big chip
without regard to the surrounding area.   If each small area
is considered in turn, the computation will clearly be not
much greater than proportional to the number of gates, but
there is a need for an earlier, global planning stage to
assign nets to channels before the detailed routing is done.
There may also be a need for backtracking to the global stage
if a particular channel proves impossible to route.
     There are two areas where further improvements can be
made within this scenario.   These are:

     (a) The use of prewired macros.   The greater the
regularity factor in the design the greater advantage can be
taken of laying out a block of gates once and copying that
block in the positions required.

     (b) Greater use of hierarchy.   There is no reason to
stop at two levels in the hierarchy of the design, global
and local, and some advantage may be gained from matching
the layout hierarchy to the conceptual hierarchy in the input text provided by the designer.   This input should also
be aimed at regularity and reduction of the number of
connections between subsystems - both qualities helpful to
auto layout - because these aims also reduce the complexity
of the system as a whole enabling a better understanding on
the part of the designer.
     An example of the text input for a hierarchical layout
system which employs channel routing is shown in Fig.12.10,
where the blocks to be laid out consist of rectangles with
connection points on their boundaries.
     In Fig.12.10 the name of each component b_1, b_2 and b_3
to be used is associated with its length and width in
brackets, followed by a list of the signal names to be found
on the North, East, South and West faces of the block.   Each
signal name is followed by its distance along the block and
the layer used to connect to it (polysilicon, metal or
diffusion.)
     A positional statement after the keyword POS describes
the relative placement and instantiation of these blocks;
thus b_1/i_1 is instance i_1 of block b_1.   The binary operators
. and ; represent the positioning of blocks above or along-
side one another, and the unary operators $>$,Λ,$<$,$-$ represent
rotation or mirroring.
     Nets are indicated by grouping signals in square brack-
ets, each signal being represented by blockname/instance/
signal name.
     The final result is a higher order block b_0 defined
under CD, which, after compilation to verify or modify the
size parameters, can be used as a definition elsewhere.
     Fig.12.11 shows the result of the auto layout of this

block where the channel widths required have been calculated
during the layout process.

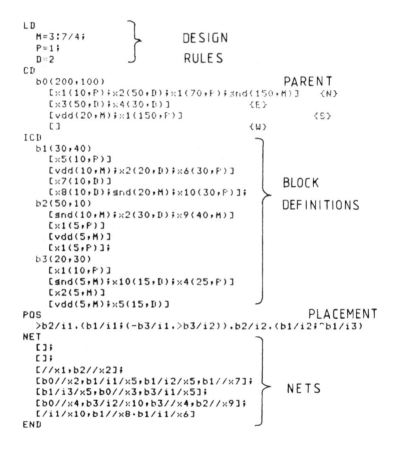

Fig.12.10   Input to a hierarchical router

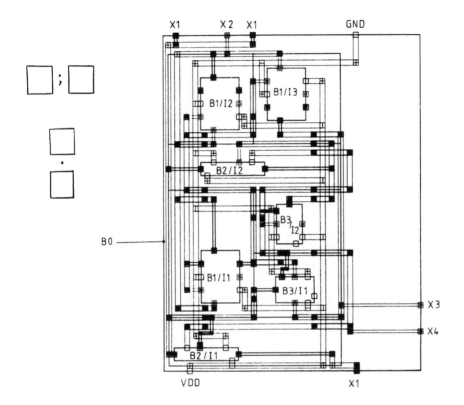

Fig.12.11   Layout produced

REFERENCES

1.   Donath, W.E., Heller, W.R., and Mikhail, W.F., 1977, 'Prediction of wiring space requirements for LSI', Proceedings 14th Design Automation Conference June 20-22, New Orleans, 32-42.

2.   Donath, W.E., and Mikhail, W.F., 1981, 'Wiring space estimation for rectangular gate arrays', VLSI 81, Academic Press, 301-310.

3.   Lee, C.Y., 1961, 'An algorithm for path connections and its applications', IEE Trans., EC10, 346-365.

4.   Aramaki, I., Kawabata, T., and Arimoto, K., 1971, 'Automation of etching pattern layout', CACM, 14, 720-730.

5.   Hashimoto, A., and Stevens, J., 1969, 'Wire routing by optimising channel assignment within large apertures', Proc. 6th Design Automation Workshop, 155-169.

Chapter 13

Design for testability

M.A. Jack

13.1 INTRODUCTION

The advantages in terms of increased complexity, improved performance, reduced costs and new systems applications made available as silicon integrated circuit technology matures from the level of large scale integration (LSI) to very large scale integration (VLSI) have been widely recognised by industry and by governments. However, one important facet of integrated circuit technology which lags dangerously behind the complexity potential of VLSI, is the problem of establishing the integrity of the design in terms of initial design validation; manufacturing quality and longer term operational reliability.

Testing of circuits with a few hundred logic functions can, in general be performed by the use of selected logic stimuli (Mueldorf and Savkav (1)). Exhaustive testing of circuits demands that all possible logic states in which a circuit can exist must be considered. Automatic test pattern generation (ATPG) methods (Williams and Parker (2), Papaionnou (3), Schnurmann et al (4)) can be used to effect in determining the test stimuli (or test vectors) required to achieve or approximate such an exhaustive test. To date, fault statistics have not permitted automatic test verification methods (Armstrong (5), Breuer (6), Szygenda and Thompson (7)) to be vigorously applied to evaluate the effectiveness of ATPG systems and the best measure of test effectiveness has involved fault simulation programs based on a simple fault model for the system under test and often using the same software as the ATPG. The inefficiency of this test scheme has been highlighted for integrated circuits of VLSI complexity and many manufacturers have adopted a scheme of 'functional testing' whereby an integrated circuit is tested by determining if all of the basic functional (and performance) requirements are fulfilled. This scheme does not however encompass previously unconsidered functions which the integrated circuit should be capable of fulfilling and which might be attempted at some point in the history of the circuit.

For combinational circuits where the present states of the output variables are a function only of the present states of the input variables, Fig. 13.1, exhaustive testing requires derivation of a test sequence to create all of the

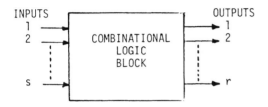

minimum number of test patterns required: 2^S

Fig.13.1 Combinational logic - exhaustive test
         requirements.

possible input combinations and check the outputs for
correct responses. These input stimuli can be applied from
automatic test equipment systems (ATE) and the responses can
subsequently be sensed by the same equipment. For a combinational circuit with s inputs a total of 2^S test vectors
would be required. For sequential circuits the present
states of the output variables are dependent on the present
states of the input variables and also on the previous
states of the circuit itself (stored as state variables or
secondary variables). Thus if the sequential circuit has a
total of m state variables, the required number of input
test vectors increases to $2^s \times 2^m$, Fig. 13.2. Clearly with

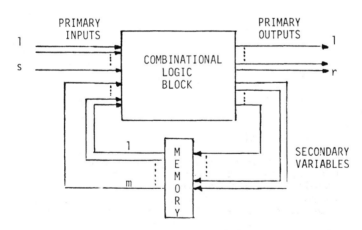

minimum number of test patterns
required: $2^{(m+s)}$

Fig.13.2 Sequential logic - exhaustive test
         requirements.

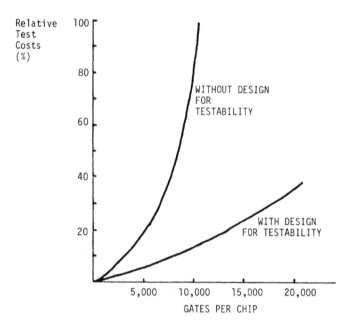

Fig.13.3 Comparison of test costs with and without
design for testability considerations
(after Grassl. (9)).

circuits of VLSI complexity, such test strategies become
uneconomical since the costs and times involved with test
pattern generation grow exponentially with increasing complexity. Minimisation of test pattern lengths becomes less
effective since the fault simulation programs - often based
on the model of a single stuck-at-fault at gate level (Williams and Parker (8)), become inefficient in terms of total
run time and in terms of the validity of the model itself.
Based on this test approach the costs of testing grow exponentially (power of 3 or 4) with increasing complexity, in
comparison to the almost linear characteristics of test
costs for circuits which incorporate a design for testability scheme (Grassl (9)), Fig. 13.3.
    This paper addresses the need to embody a testability
scheme within the VLSI integrated circuit design itself and
catalogues several techniques which constitute the term-
design for testability. The techniques considered are generally applicable in logic design, ranging from testability
schemes which partition (Hayes (10)) the design into manageable blocks for test purposes, possibly using the system bus
for data access; to schemes which permit self-test of the
circuit as it operates, and in some cases allow correction
of errors (Peterson and Weldon (11)); to schemes which permit all of the test 'equipment' to be included on the silicon as built-in test. Built-in test differs from self-test
in that with built-in test schemes an external initiation
control must be used. The specific built-in test schemes
considered are the scan-path technique (Williams and Angell

(12), Das Gupta et al (13)), the signature analysis technique (Nadig (14), Frohwerk (15)) and the extension of these concepts to the built-in logic block observation (BILBO) method (Koenemann et al (16)).

Whereas, in the past, the driving design consideration for LSI was minimisation of active silicon area (maximisation of production yield), the future criterion will become predominantly testability.

13.2 DESIGN FOR TESTABILITY

Two key concepts pervade all strategies of design for testability: these are controllability - the ability to establish the circuit in a controlled initial state, and observability - the ability to observe externally, the internal circuit states (Goldstein (17)). A typical integrated circuit consists basically of a series of combinational logic blocks with associated clocked memory elements operating as a synchronous sequential circuit. As the number of inputs, outputs and state variables increases, the test requirements increase since the circuit contains more gates resulting in a linear growth of test costs (linear growth in possible fault sites). The logical depth (the number of gates involved in logic determination) and the sequential depth (the total number of possible states in which the logic can be placed) both increase, causing an exponential rise in test costs. Thus increased logical complexity has the consequence that the controllability and observability of the tests are adversely affected since not all of the possible logic states may appear at the circuit pins and in any case, since the length of the input (stimulus) sequence and the length of the output (response) sequence increase exponentially, it may not be possible to check the logic states in reasonable timescale or cost limits. Analytic measures of controllability and observability in sequential networks have been reported by Goldstein (17) or Kovijanic (18) and these offer pointers to those critical parts of a design which must be incorporated in the design for testability strategy.

Design for testability involves increasing the controllability and observability of the constituents of a design by decomposing the overall design into more manageable elements for test purposes.

13.3 TEST STRATEGY REQUIREMENTS IN VLSI DESIGN

An integrated circuit test strategy must ideally allow for a range of differing test environments to be experienced by the circuit during its operational service. These environments can be summarised as:
(a) prototype characterisation; to include design validation and parametric testing.
(b) production test; to include yield enhancement.
(c) service or maintenance test; to include self-repair.

In prototype characterisation it is essential to identify and localise individual faults to enable fault diagnosis and correction. Prototype faults may be process-related faults statistically distributed over a processed wafer, or they may be design faults (errors) such as forgotten contact holes, wrong interconnections or excessive signal delays. Prototype testing is invariably carried out by the designer(s) using ATE, microprobing or electron beam facilities.

Production test requirements include both process quality checks and functional checks. Process quality control is achieved either by a number of chip-size 'drop-in' replacements spaced over the wafer or by using a small test area on each chip. Measures of transistor parameters, contact resistance and capacitance values are made to check production tolerances. In production test, functional (and parametric) tests must be minimised since here testing time and costs are important. Functional tests need only yield a limited number of the significant internal states since it is not generally possible to redesign or repair at this stage.

In maintenance and systems test, fault diagnosis is precluded so a simple GO/NO-GO indication for the circuit is adequate.

It will be shown that a number of the design for testability strategies considered here offer the potential of valid use at each stage in the life of a VLSI circuit. In addition, use of circuit design techniques which eliminate dynamic faults with synchronous logic and which minimise the functional complexity of blocks of circuitry, assist in improving the overall efficiency of a test strategy.

The freedoms available to achieve overall testability requirements are bounded by three considerations:

(a) the additional number of package pins available for test
(b) the need to include as few additional circuits on chip as possible
(c) the need to minimise loss in chip performance.

The number of package pins is an important factor since larger packages are more expensive in both basic costs and assembly time. Additional test pins require more silicon area (pads) and more power dissipation. However, pins required for testing can 'steal' access via other pin logic combinations which might not appear in normal operation. Alternatively it may be possible to use multiplexer circuits for test input/output via a small number of pins. Use of a CLEAR or PRESET in the circuit structure will offer a 'free' test control line.

Increased circuit complexity reduces production yield thus the increased chip costs involved in using extra silicon area for test purposes must be weighed against the savings in testing costs (time). Typically, use of test circuits which increase chip size by 5-10% is considered reasonable, Fig. 13.4.

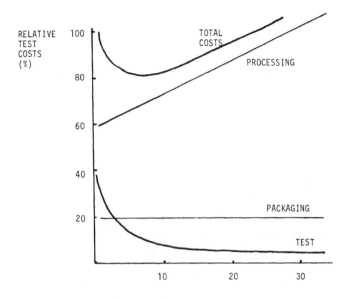

TEST CIRCUIT AREA OVERHEAD (%)

Fig. 13.4 Typical variation of chip costs as a
function of test circuit area overhead
(after Grassl. (9)).

Design for testability schemes may result in a reduction in circuit performance since higher gate loads or increased logic delays may be involved. Such degradations may
demand modification to the overall system architecture.
For example, pipelining may become necessary to achieve performance specification.
The overall significance of the test area overhead
depends on the type of chip being designed. For low cost,
high volume, modest performance designs (microprocessor) an
acceptable test overhead is 5-10% whereas for high performance, low volume applications (military) test overheads of
the order of 100% may be acceptable.

13.4 PARTITIONING FOR TESTABILITY

An integrated circuit is normally composed of several
logic blocks, therefore partitioning of the circuit into
these blocks for test purposes is a convenient method of
reducing the test problem. Since an exponential relationship exists between test time and circuit complexity, the
total time required to test the individual logic blocks will
be significantly less than the time required to test the
chip as an entity. An acceptable level of partitioning can
be estimated from the improved efficiency of ATPG systems
achieved by reducing the partitioned constituents to minimal
complexity, or in the limit to purely combinational elements. Circuit partitioning for test purposes is a very

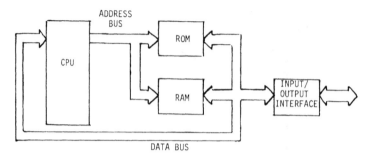

Fig.13.5 Circuit partitioning in bus-oriented
designs.

realistic test scheme since, in general the circuit will be
composed of elements produced by different designers. Thus
if the design of one logic block is changed, only the testing for that specific block must change.

In a bus-structured design (microprocessor) most of
the logic blocks will already access the bus and normally
the bus will be externally accessible, Fig. 13.5. The bus
therefore naturally partitions the VLSI circuit into smaller
blocks and offers a direct path to the blocks for improved
controllability and observability. One drawback of this bus
driven test strategy exists for faults on the bus itself.
If a bus line exhibits a stuck-at-fault this may be produced
by any of the logic blocks or by the bus itself. Under
these conditions, use of scanning electron microscope (SEM)

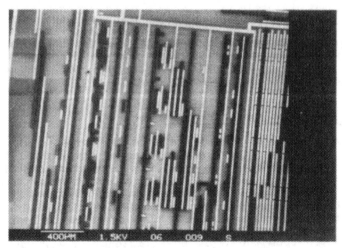

Fig.13.6 Voltage contrast scanning electron
microscope photograph of part of a
circuit.

techniques, possibly with voltage contrast facilities can be of advantage.   Fig. 13.6 shows a section of a circuit under test in such a voltage contrast scanning electron micro- scope.   Here the aluminium interconnections show as light regions or dark regions depending on their electrical poten- tial (0V = LIGHT; +5V = DARK).   In effect the voltage con- trast technique presents the designer with a 'map' of the logic states on the bus and affecting the bus, thus facili- tating fault isolation.

13.5 SELF-TEST

Self-test schemes for integrated circuits involve the incorporation of extra test circuitry on-chip such that this test circuitry forms an adjunct to the computation processes of the chip.   As such the self-test circuitry is continual- ly active over the working life of the chip, yielding error detection information or in some cases, providing error correction.   Such self-test schemes, by definition require use of otherwise redundant (duplicated) circuitry for error detection and correction.   The test overhead involved in designing double or triple redundancy with associated (ma- jority) voting circuits for self-test in a VLSI chip, Fig. 13. 7, limits the use of such redundancy techniques to critical parts of a design.   Similarly the test overhead involved in use of error correcting codes, Fig. 13.8 usually limits the use of such codes since a 16-bit word requires an additional 6 parity (check) bits for correction of a single error and

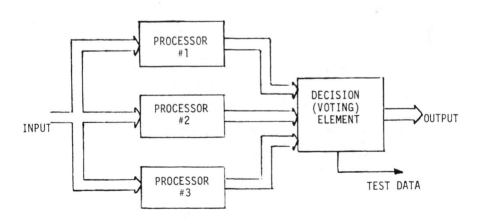

Fig. 13.7 Self-test scheme based on triple
      redundancy.

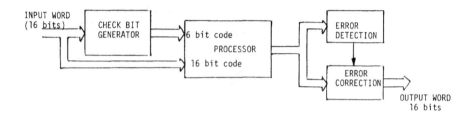

Fig.13.8 Self-test scheme based on error
          correcting codes.

detection of a double error (Peterson and Weldon (11)).
Test overheads in such a scheme would be of the order of
30%. Single error detection for such a 16-bit word needs
only 1 parity bit with an acceptable 6% test overhead.

As an alternative to the use of silicon area directly
for self-test, it is possible to exploit the inherent intel-
ligent or computational attributes of certain classes of
design (microprocessors) to monitor results and indicate the
validity of these results.

13.6 BUILT-IN TEST

Built-in test differs from the self-test schemes considered in the previous section in that built-in test
schemes demand external activation to inhibit the normal
operation of the VLSI circuit and permit the test sequence
to become active. Built-in test effectively incorporates
(to various degrees) the test equipment for both stimulus
(controllability) and response monitoring (observability) as
part of the design, Fig. 13.9.

Possibly the simplest example of built-in test is the
strategy of test by stored microprogram used in
microprocessor-type structures, Fig. 13.10,where a 'shadow'
processor can be included for built-in test purposes by including some extra microcode instructions and some additional read-only-memory (ROM) which is activated during test to
generate pre-determined stimuli to the circuit under test.
Here the computational attributes of the microprocessor can
be used to advantage during the testing operations. Thus
the possibility of using the microprocessor to test itself
exists (Boney (19)) thereby minimising or eliminating the
need for conventional test equipment.

A more general approach to built-in test involves the
use of scan-path techniques which permit a significant improvement in the controllability and observability measures
for a VLSI design by connecting critical circuit nodes to a
long shift register which permeates through the circuit. A
range of scan path techniques known variously as level-
sensitive scan design (LSSD), scan path or set/scan logic
have been reported however all exploit similar concepts, the
main difference being in the design method employed for the

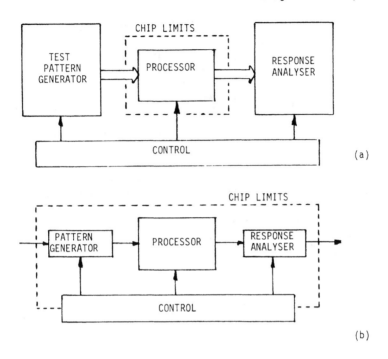

Fig.13.9 Basic form of built-in test scheme.

      (a) normal test arrangement with
          external test equipment.

      (b) test equipment included in
          circuit design.

scan path circuits to eliminate race and dynamic hazards and
improve the testability of the scan-path itself.

13.7 THE SCAN PATH

The scan-path enhances the controllability of a VLSI
design since it allows the serial-in/parallel-out shift register of the scan-path to define circuit states prior to
test, Fig. 13.11(a). Similarly the scan-path enhances the
observability of the design by allowing critical circuit
states to be accessed by the (now reconfigured) parallel-
in/serial-out shift register scan-path, Fig. 13.11(b). The
simplest form of scan-path test strategy is to include, on-
chip, an additional shift register with connections to/from
various nodes on the chip. Clearly this involves increased
circuit area. A more attractive scan-path arrangement
would involve functional conversion of existing flip-flops,
shift registers or gates to the required scan-path shift register configuration using multiplexed control signals and

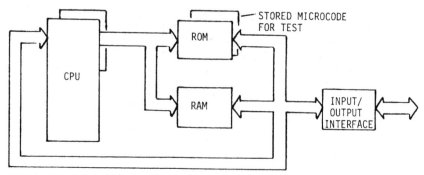

Fig.13.10 Built-in test by stored microprogram.

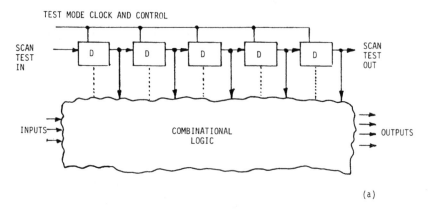

(a)

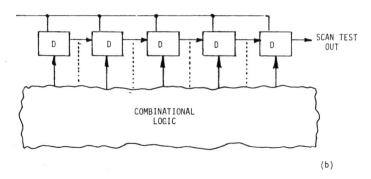

(b)

Fig.13.11 Built-in test using a scan-path.

     (a) scan-path as serial-in/parallel-out register to control initial circuit states.

     (b) scan-path as parallel-out/serial-in register to observe final circuit states.

thereby reducing test area overheads to a minimum. The method of testing using a scan-path is as follows. Firstly the scan-path itself is tested by entering a sequence at the input and monitoring the sequence obtained at the output for correctness. Secondly, the circuitry present between nodes attached to the scan-path can be tested by shifting-in a pre-determined test sequence and latching the bits of the test sequence to the appropriate test nodes. The circuit can then be switched to normal operation while a set of test vectors is applied to the primary inputs of the chip. The primary outputs of the chip may be monitored at this time. Finally, the internal circuit states can be loaded into the shift register and shifted along the scan-path to the output while a new test sequence is being fed-in.

The total test time is determined mainly by the number of stages in the scan-path which, in turn, is defined by the number of individual blocks of logic which are to be tested. Optimum test requires inclusion of a complete scan-path which leaves no sequential logic circuits during the test mode. However speed, performance or area constraints may restrict the use of the scan-path to an incomplete scan-path which leaves parts of the circuit which are sequential and possibly not exhaustively tested.

The test overhead involved with the scan-path strategy of built-in test depends on the basic form of the circuit structure and the availability of circuit elements suited to functional conversion. The effect of circuit performance is only of importance if additional scan-path register stages have to be included in the design. Otherwise, only increased delays due to loading and routing delays may occur.

The primary advantage of the scan-path method lies in the fact that only 3 extra circuit pins are required for test enable and data. However, the scan-path merely allows access to circuit nodes and use of the scan-path still demands external pattern generation and response monitoring equipment to derive the required test results.

13.8 SIGNATURE ANALYSIS

The scan-path provides excellent controllability and observability for a circuit but much of the actual test time is wasted in serial transmission between the circuit and the (external) test equipment. An alternative test scheme is to build the required test pattern generators and state monitoring circuits into the VLSI design. As discussed previously, for designs with computational capabilities (microprocessors) it is often convenient to achieve such built-in test schemes by extra storage to enable test by stored micro-program. However, a technique with wider application involves on-chip generation of a pseudo-random binary sequence (PRBS), with associated signature analysis of the resulting logic pattern at some point in the circuit.

Pseudo-random binary test sequences are well-suited to on-chip generation since they can be produced very easily by

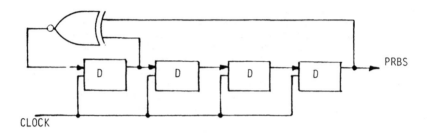

Fig.13.12 Basic form of PRBS generator.

shift register circuits of the form shown in Fig. 13.12
Further, PRBS generators are well suited to functional
conversion of existing circuitry and minimisation of test
overhead can be achieved in certain types of circuitry. If
the feedback taps of the PRBS generator, Fig. 13.12, are
well-chosen (Peterson and Weldon (11)) an n-stage shift register can generate a pseudo-random bit sequence of length 2^n
-1 bits.

Monitoring the test response can be performed using a
signature analysis register, Fig. 13.13 which is basically a
PRBS register with extra input(s) to the feedback control
circuit. The signature analysis sequence proceeds as follows. Firstly, all inputs to the logic block under test
are initialised to a predetermined pattern, possibly one
convenient state of a PRBS generator. Then the circuit is
allowed to operate, feeding the output from the node(s)
under test to the signature analysis register. After a
predetermined number of system clock cycles, the signature
of the node under test is compared with the expected signature and a GO/NO-GO indication can be given.

The example of Fig. 13.13 shows that after 10 cycles
which produce the node sequence shown, the signature
analysis register should be in the state <1001> and this
signature can be compared to the actual signature produced.
Any error in the node sequence will cause the signature to
rapidly diverge from the expected final state and, in general, the probability of detecting an error in a node stream
of length m bits with a signature analysis register of
length n is (Nadig (14), Frohwerk (15))

$$P(m,n) = 1 - \frac{[2^{m-n}-1]}{[2^m-1]}$$

In Fig. 13.13 the probability of detection is 0.92.
Since the test overhead involved in formation of both
PRBS generation and signature analysis registers is low, and
functional conversion of existing circuitry is possible,
many such signature analysis test schemes can be incorporated in a design. The tests can then proceed simultaneously
with separate logic blocks having their own signature

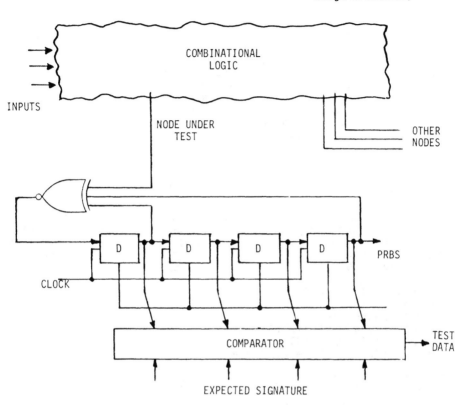

SIGNATURE REGISTER STATES				NODE UNDER TEST
0	0	0	0	1
0	0	0	0	1
0	0	0	0	1
0	0	0	0	0
1	0	0	0	1
1	1	0	0	1
1	1	1	0	0
0	1	1	1	0
0	0	1	1	1
EXPECTED SIGNATURE 1	0	0	1	

Fig.13.13 Built-in test by signature analysis.

analysis register. Alternatively, a reduced form of signature can be employed where exclusive NOR gates each fed from a node in the logic block, are placed between register stages. The initialisation pattern for the logic block can conveniently be chosen such that the (correct) anticipated signature is the all-zero state, which can be easily detected. The GO/NO-GO results from each signature analysis register can then be transmitted to a status register for read-out to test equipment or to a service processor. Note that PRBS and signature analysis techniques do not offer fault diagnostics.

13.9 BUILT-IN LOGIC BLOCK OBSERVATION (BILBO)

The BILBO test scheme merges the scan-path concept with the PRBS/signature analysis concept and since both are implemented with identical (shift register) elements the BILBO approach offers very low test overheads. Each BILBO is required to be configurable in 4 main forms:

(a) serial-in/parallel-out shift register (scan path) to control initial states.
(b) PRBS generator.
(c) signature analysis register.
(d) parallel-in/serial-out shift register (scan-path) to observe final signature or signature analysis decision.

The BILBO elements are placed in a design to isolate purely combinational blocks, Fig. 13.14, and during test, pairs of BILBO elements operate as a PRBS generator/signature analysis register and the logic blocks can be tested consecutively. At the conclusion of the tests the signatures can be output with the BILBO elements in scan-path mode. The BILBO technique thus exploits the efficient test pattern generation and fault detection properties of signature analysis whilst minimising the output test data volume to a series of test signatures or GO/NO-GO decisions for each logic block.

13.10 TESTABILITY IN SEMI-CUSTOM CIRCUITS

In most semi-custom designs the basic scheme of scan-in/scan-out is employed to enhance testability aspects (Clifford (21)) however the inherently "standard" block nature of cell-based approaches and the inherently repetitive nature of gate arrays tends to increase the test area overhead to around 10%-20% thereby affecting the overall gate utilisation for logic operation and it has been suggested (Hoshikawa et al (20)) that dedicated test circuits should be included to enhance the utility of scan-path methods.
The drive towards the inclusion of testability structures in semi-custom circuits is two-fold. Firstly, the need for "first-time-correctness" in semi-custom has placed demands on design validation and test pattern

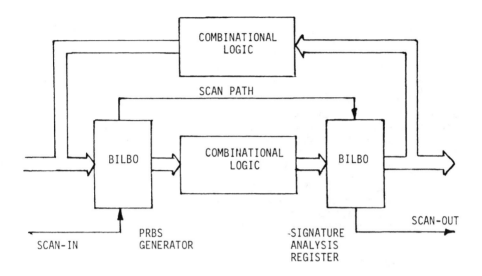

Fig.13.14 Built-in test by BILBO.

generation/fault simulation software. Fault simulation involves simulating the signal values at specified points in
the circuit for both a fault-free version and the faulty
version. When mismatch occurs the error is deemed to be
detected.

Secondly, the (stuck-at) models normally employed in
such simulations have been shown to be inadequate, particularly for dynamic circuits and CMOS technology in particular.

Current trends are towards the incorporation of "tran-
sparent" testability structures within standard cells or
within functional groups (macros) for gate arrays where
these structures can be activated by means of multiplexer
lines. Associated with these structures are peripheral
test elements such as PRBS generators or signature analysis

registers.

13.11 CONCLUSIONS

The need to incorporate within integrated circuit
design methods, a strategy to enhance testability has been
stressed and it has been shown that attempts to test circuits of VLSI complexity using conventional
stimulus/response methods are totally inefficient. Attempts
to improve the efficiency of this approach to testing by use
of automatic test pattern generation and fault simulation

programs and high performance (functional) test systems offer very restricted potential for future improvements.

The general utility of built-in strategies has been discussed and the functional simplicity of scan-path and signature analysis schemes has been stressed. Such built-in test strategies offer high efficiency of test throughout the life of the chip where BILBO test schemes (incorporating scan-path and signature analysis concepts) can be modified to accommodate the different test environments encountered by the chip. In design validation the scan-path features of the BILBO permit fault isolation; in production test the PRBS/signature analysis features of the BILBO permit rapid test while the scan-path facility enables the test data to be accessed externally. In the service/maintenance environment the BILBO scheme can yield an overall signature analysis GO/NO-GO decision for the VLSI circuit.

Testing of VLSI systems cannot be allowed to remain as an afterthought to design and processing. Testability must be considered at all levels of the design operation and new VLSI design methodologies must exploit testability considerations to ensure acceptance.

REFERENCES

1. Mueldorf, E.I. and Savkar, A.D., 1981, "LSI logic testing - an overview", IEEE Trans. Comput. Vol.C-30, pp.1-17.

2. Williams, T.W. and Parker, K.P., 1979, "Testing logic networks and design for testability", Computer pp.9-21.

3. Papaionnou, S.G., 1977, "Optimal test generation in combinational networks by pseudo-Boolean programming", IEEE Trans. Comput. Vol.C-26, pp.553-560.

4. Schnurmann, H.D., et al, 1975, "The weighted random test pattern generation", IEEE Trans. Comput. Vol.C-24, pp.695-700.

5. Armstrong, D.B., 1972, "A deductive method for simulating faults in logic circuits", IEEE Trans. Comput. Vol.C-22, pp.464-471.

6. Breuer, M.A., 1970, "Functional partitioning and simulatiⁱ of digital circuits", IEEE Trans. Comput. Vol.C-19, pp.1038-1046.

7. Szygenda, S.A. and Thompson, E.W., 1976, "Modelling and digital simulation for design verification diagnosis", IEEE Trans. Comput. Vol.C-25, pp.1242-1253.

8. Williams, T.W. and Parker, K.P., 1982, "Design for testability - a survey", IEEE Trans. Comput. Vol.C-31, pp.2-15.

9. Grassl, G., 1980, "Design for testability", Digest NATO Advanced Study Course in VLSI design.

10. Hayes, J.P., 1974, "On modifying logic networks to improve their diagnosability", IEEE Trans. Comput. Vol.C-23, pp.56-62.

11. Peterson, W.W. and Weldon, E.J., 1972, "Error correcting codes", Cambridge MA, MIT Press.

12. Williams, M.J.Y. and Angell, J.B., 1973, "Enhancing testability of large scale integrated circuits via test points and additional logic", IEEE Trans. Comput. Vol.C-22, pp.46-60.

13. DasGupta, S. et al, 1978, "LSI chip design for testability", Digest Int. Solid State Circuits Conference, San Franscisco, CA, pp.216-217.

14. Nadig, H.J., 1977, "Signature analysis - concepts, examples and guidelines", Hewlett Packard Journal, pp.15-21.

15. Frohwerk, R.A., 1977, "Signature analysis: a new digital field service method", Hewlett Packard Journal, pp.2-8.

16. Koenemann, B., et al, 1979, "Built-in logic block observation techniques", Test Conf. pp.37-41, IEEE-79CH1509-9C.

17. Goldstein, L.H., 1979, "Controllability/observability analysis of digital circuits", IEEE Trans. Circuits Syst. Vol.CAS-26, pp.685-693.

18. Kovijanic, P.G., 1979, "Testability analysis", Digest Test Conf. pp.310-316, IEEE-79CH1509-9C.

19. Boney, J., 1979, "Let your microprocessor check itself and cut down your test overhead", Electron. Des. Vol.18.

20. Hoshikawa, R., et al, 1982, "Introduction of an ultra fast 800-gate CMOS gate array", Proc. 2nd Int. Conf. on Semi-Custom Circuits.

21. Clifford, C., 1982, "Gate array testing philosophies", Proc. 2nd Int. Conf. on Semi-Custom Circuits.

Chapter 14

Silicon compilers and VLSI

Peter B. Denyer

14.1 INTRODUCTION

For some considerable period custom LSI and VLSI applications have been technology led. Relentless advances in fabrication technology result in ever more potential for new designs. Chips containing 100,000 transistors are now available, and million transistor devices already appear feasible. Yet designers are struggling to fulfill this potential because of the enormous complexity of their task, which appears to grow faster than linearly with circuit size.

To a significant extent the semi-custom techniques discussed in this book have helped to keep the design task manageable. However, continuing process improvements force us to consider yet more powerful techniques and tools to keep pace with this growth in complexity.

The structured approach to tackling complex systems is well recognised as desirable, if not essential, for the ultimate development of a successful design. This involves a naturally hierarchical structure of design, like the pyramid shown in Figure 14.1. The top of the pyramid contains the highest level of system description and the design process involves successively splitting this task into independent operations on ever-lower levels. At the bottom of the pyramid are found myriad independent layout, simulation and checking tasks at the lowest cell level. The process of implementation involves executing these small, manageable tasks and then combining their results via rules of composition to progress back up the pyramid and so realise the entire system.

It is helpful to consider the impact of Standard-Cell and Gate Array techniques in the context of this 'pyramid of tasks'. The Standard-Cell approach obviates the design and layout of the lowest (gate) level cells. This can account for a substantial proportion of the bottom of the 'pyramid' and so may dramatically reduce the total task. Similarly, in their most primitive form, Gate Arrays relieve us from the task of defining and placing transistor geometries. This at least removes the very bottom layer of the pyramid. However, we might also use a library of 'standard-cell' macros on the Gate-Array to make commonly required gate-level elements. Again this offers the same advantage as the

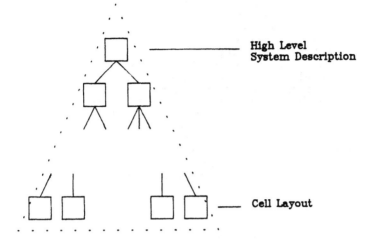

Figure 14.1   A pyramid of tasks generated by a
structured approach to design.

Standard-Cell approach.   Indeed, the Gate Array is now to
be seen simply as an alternative target floorplan, albeit
one with special advantages and disadvantages.
    Whilst the majority of this book is devoted to the
power of such techniques, they in turn begin to appear
inadequate as chips become more complex.   In effect, the pyramid is pushed further and further up into the system,
eventually (perhaps) encompassing all of it.   Certainly,
help on the lower levels is still attractive, indeed essential, but the remainder of the complexity pyramid is now
perhaps as large as the one we were faced with in the first
place.
    Clearly, VLSI technology forces us to consider yet
more powerful semi-custom techniques to keep system design
within the grasp of a realistically large user community.
The remainder of this chapter discusses some of these techniques, as they are in use today, and as they are perceived
to develop in the future.   It may come as no surprise to
find that they are all forms of computer <u>automated</u> design
(as opposed to computer <u>aided</u> design), for it is clear from
the above discussion that the design problem is rapidly out-
growing ordinary human capability.

14.2   AUTOMATIC PLACEMENT AND ROUTING

    The next level in the complexity pyramid concerns the
physical placement and wiring together of gate-level elements from the cell library.   With total freedom in two dimensions, efficient solutions to this problem become chal-
lenging research tasks.   The problem is more easily defined
however for the restricted chip architectures typically associated with Gate Array and Standard Cell designs.

Gate Arrays pose the hardest problem, for the chip topology is totally predetermined and wires must be routed
within fixed channel widths.   Of the known routing algorithms,   the   'line-search'   technique   [1]   is   popular   for
closely packed 'grid' arrays.   As shown in Figure 14.2 this
routing algorithm respects obstructions (e.g. previously
filled channels) by constructing a framework of legal path
possibilities and then determining a realistic connection.
Typically a line-search algorithm may route a 1000-gate array in the order of five minutes.   Within such restrictive
conditions however, 100% completion is not generally expected.   The degree of success is linked to the percentage site
utilisation.   For example, a circuit using 80% of the active array sites might route (say) 95% successfully.   The
human designer is left to complete the task!

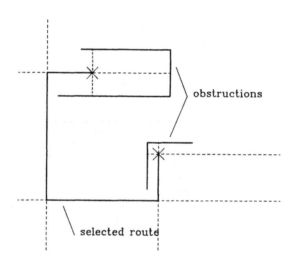

Figure 14.2   An example of the Line Search technique
for automatic routing.

Alternative routing algorithms are useful for regi-
mented Gate Array or Standard Cell topologies.   In this case
the active circuitry is packed in rows of uniform height,
with channels left between the rows for routing.   Again the
channel width is fixed for the Gate Array, but may be varied
to suit demand for Standard Cell designs, which are not
pre-processed. The 'channel routing' task is relatively
straightforward,   and   where   the   channel   width   is   not
predetermined (Standard Cell) can be 100% successful.   Even
for fixed channel widths 100% completion is feasible if the
interconnect channels are wide enough [2].   It has been argued however that arrays designed with this capability use
chip area inefficiently.
In any case, where the routing is not entirely completed automatically several possibilities remain to com-

plete the design;

* manual rearrangement

* automatic removal and rewiring in congested areas

* reorder cell placement and start again.

The latter option is attractive if a fast routing algorithm
is available.

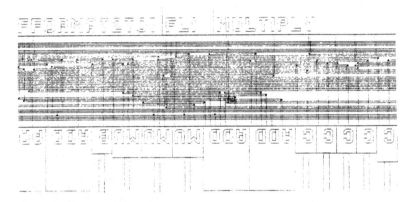

Figure 14.3   An example of channel routing (taken
from the FIRST silicon compiler).

14.3   AUTOMATED CUSTOM DESIGN

It is not our purpose here to examine the many significant advances in techniques and design aids for fully custom VLSI.   However, when these reach the point that the
custom commitment is entirely automated, then they may be
argued to offer a semicustom capability.   Certainly if the
end result permits silicon circuit design without involvement at the transistor level, then the technique offers the
same advantages as the Standard Cell and Gate Array approaches.   As we shall see, these techniques may be viewed
as a next-generation Standard Cell approach.

Suppose an expert custom designer has completed the
design of an 8-bit adder.   Requested to attempt a 12-bit
adder, he is likely to complete this new task quite rapidly.
He will use cells from the 8-bit adder, and apply some
'rules' for adder construction (in this case, perhaps as
simple as stacking the cells side by side).

Given an appropriate programming environment [3], it
is possible that these rules could be automated, allowing
adders to be arbitrarily generated from a library of
predesigned cells.   Note an important difference here over
the Standard Cell approach.   Module generators are one level
up from the cell-library.   Instead of calling cells,

designers call procedures which in turn call and combine cells to make modules.

The adder is a good example, but perhaps the most pre-valent module generators are the PLA synthesizers that are now found at many design sites.   Figure 14.4 is a typical example of an automatically generated module, in this case a bit-serial multiplier made from five cell types.   Cell C is a beginning cell, Cell D is an end cell, Cell E is an inter-mediate cell and Cells B and H are interconnect.   These multiplier cells work two bits at a time, so calling a 12-bit multiplier involves the assembly procedure for the be-ginning and end cells, four intermediate cells and the in-terconnect cell.   Figure 14.4(a) illustrates the assembly procedure while 14.4(b) shows the full layout generated when each of the cell calls is instantiated.

The message here should be clear, that for a little software investment, designers may greatly improve their fu-ture productivity.   For frequently used modules the return on this investment can be enormous.   As a further multiply-ing effect, the skills of the original module designer are effectively transmitted to an entire design community.

(MULTH)		
(MULTD)	(MULTB)	(MULTD)
(MULTD)	(MULTB)	(MULTD)
(MULTC)	(MULTB)	(MULTE)

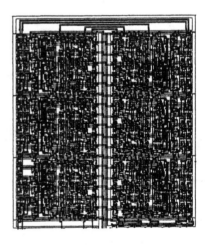

Figure 14.4   An example of automatic module generation by procedure.
(a)   Cell calls and placement.
(b)   Instantiated layout.

14.4   CORRECTNESS BY CONSTRUCTION

A major advantage of all automated routing and circuit synthesis is the associated value of correctness.   Once the cells and algorithms themselves are proven correct then the circuits produced by them can be assumed to be correct every time.   This concept is more powerful that it at first ap-pears, for although automated synthesis is convenient in its own right, at the same time it obviates any 'checking' requirements. Indeed the saving in this area can be remarkable.

For example, to check that the mask layout faithfully corresponds to the desired electrical network a range of 'circuit extraction' tools are now available to custom designers. Typically these may take 10 hours to run on a 20,000 transistor circuit. This process must be repeated every time there may be a chance of error (i.e. after <u>any</u> human intervention). It may only be avoided where the synthesis has been entirely automatic. Similar arguments apply to other essential checks, such as Design Rule Checking (DRC).

Thus automated synthesis techniques appear to offer a double advantage over hand-designed layout and routing.

14.5   CHIPS AS MODULES

Instead of calling procedures that synthesize modules, perhaps we could call procedures that would synthesize entire chips. Actually, in a suitable environment any chip design can be defined procedurally, like a module, and it is likely that with care we might be able to change some parameters and 'reassemble' the chip automatically.

This obvious extension of the module-generator is interesting, but rather narrow since we will need a detailed cell composition 'procedure' for every chip to be built. Clearly some higher level facility would be desirable.

14.6   THE SILICON COMPILER

Consider now a tool which will take over the entire design hierarchy, automatically synthesising entire VLSI chips (or systems) from some high level programmatic description. Such a tool is in effect a 'Compiler', except that its ultimate result is no longer machine, or assembly language code, but detailed layout geometry.

The distinction is a powerful one, and the tool itself more so, since it would appear to replace many man-years of expensive design effort. Not surprisingly the concept has been maligned by many experienced hands, but since its introduction by Johannsen [4] there has been much development, and infant compilers of significant power are already appearing (e.g. MacPitts [5], FIRST [6]).

The Silicon Compiler then is an ultimate semi-custom tool, taking a high-level system description as input, and supporting all of the electrical and geometrical synthesis processes.

Normally these will be entirely automatic, with no human intervention. Typical of those compilers so far reported is the five-step process shown in Figure 14.5, as used in FIRST.

The input language supports functional description in some form and also a subset of the normal programming con-

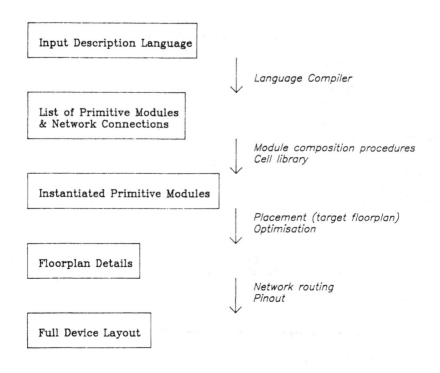

Figure 14.5   Five steps to silicon compilation.

ventions (e.g. integer arithmetic, procedural definition). These powerful features can dramatically affect the amount of code required to define a system.

The first step is to 'compile' this programme, obeying the procedures and evaluating the expressions to produce an intermediate list of 'primitive' modules and an interconnection network. Primitives are the low level instruction-set of the compiler, they have a physical representation (layout) which is held as a library of cells and composition procedures. Thus primitives are typically modules in our previous terminology. Examples from FIRST include MULTIPLY, ADD, DELAY, etc.

The entire list of primitives is then instantiated. To form the complete chip these must be placed and connected, again automatically. The placement process is more easily tackled when there is a target physical architecture, or floorplan. This floorplan is intimately linked to the system architecture and must be carefully chosen to ensure acceptable layout efficiency and also to offer a suitable framework for communication. MacPitts supports a data-path floorplan style for bit-parallel register-based architectures. FIRST is intended for bit-serial architectures and supports a 'Manhattan Skyline' floorplan (see Figure 14.6) in which the entire system network is routed in a central communication core. In either case chip areas are affected

by the order of placement and both FIRST and MacPitts attempt to choose an optimal order in this respect.

The chip is completed by adding appropriate pad and
peripheral utilities and automatically routing the network.

Both the cell library and the routing are effectively
technology bound, but higher levels in the Compiler are not.
This leads to the interesting proposition that the initial
system description is <u>technology independent</u>. Herein lies a
powerful advantage to system design and documentation.

Figure 14.7 tells the whole story. Shown in the
first part of this figure is a small input file to FIRST
describing the signal flow graph of a particular filter system. The second part of this figure shows an entire chip
layout produced automatically by the Compiler from the same
input file. Among other definitions that describe the Silicon Compiler, this ratio of 20 lines of code input to 4,000
transistors output is perhaps the most graphic.

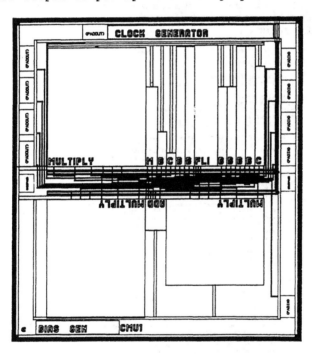

Figure 14.6   Example of a target floorplan archi-
              tecture (Manhattan Skyline).

14.7   <u>ADVANTAGES</u> <u>OF</u> <u>SILICON</u> <u>COMPILATION</u> <u>AS</u> <u>A</u> <u>SEMI-CUSTOM</u>
TECHNIQUE

Despite the brevity of the previous descriptions it
should be clear that the Silicon Compiler is potentially an
extremely powerful semicustom technique. In the first in-

stance the user is not required to be experienced in IC design; the Compiler is the 'expert'. More than any of the previously described semi-custom techniques this opens the door to VLSI silicon as a flexible development medium for many system designers. Moreover, the compiler is extremely fast (typically only a few cpu-seconds for a FIRST-compiled chip). In its turn this encourages high-level experimentation in system design: several chip designs can be attempted for little real cost before final commitment. Eleventh-hour system changes (e.g. changing a 12-bit word to 14-bits) are far from catastrophic.

It is difficult to assess typical system design times, and easy to underestimate them. A conservative figure would probably be four man-weeks but, as with any powerful tool, skill and dedication can yield exceptional results. Both MacPitts and FIRST have yielded remarkable design cycles as low as two or three man-days.

On top of these advantages lies the reassurance of correctness-by-construction that is now extended to the entire VLSI system.

```
        FIRST COMPILER   - Copyright Denyer,Renshaw,Bergmann - 1982

SOURCE FILE: LMSFIR1

!Adaptive(LMS)Transversal(FIR)Filter
!filter stage of multiplexed filter sections

CONSTANT round=1
CONTROL INPUT c0,c1,init
CONTROL INPUT c1t0,c1t1,c1t2,c1t3,c1t5,c1t6,c2t0
INPUT Xin,TMEIN,WMSBin,WLSBin
OUTPUT Xout,YMSBout,YISBout,YLSBout

    OPERATOR FIR[wordlength,idles,stages]

        SIGNAL s1,s2,s3,s4,s5,s6,s7,s8,s9,s10,s11,s12,s13,s14,s15,
            s16,s19,s20,s21,s22,ss1,ss2,ss3,ss4,ss5,ss6

        Multiplex[1,0,0] (c2t0) s3,Xin -> s1
        Multiplex[1,0,0] (INIT) s15,WMSBin -> s7
        Multiplex[1,0,0] (INIT) s16,WLSBin -> s8
        Multiplex[1,0,0] (c1t2) s6,s5 -> s6
        Multiplex[1,0,0] (c1t6) YMSBout,YISBout -> YMSBout
        Multiply[round,wordlength-2,0,0] (c1t0->nc) s2,TMEin -> s5,nc
        Multiply[round,wordlength-2,0,0] (c1t5->nc) s22,s15 -> YLSBout,YISBout
        Add[1,0,0,0] (c1t3) s6,s11,gnd -> s15,nc
        Add[1,0,0,0] (c1t3) s5,s12,gnd -> s16,nc
        constant p1=(idles+1)*(wordlength)
        Bitdelay[18] s1 -> ss1
        Bitdelay[18] ss1 -> ss2
        Bitdelay[18] ss2 -> ss3
        Bitdelay[18] ss3 -> ss4
        Bitdelay[18] ss4 -> ss5
        constant p2 = p1-90 , p3 = p2/2 , p4 = p2-p3
        Bitdelay[p3] ss5 -> ss6
        Bitdelay[p4] ss6 -> Xout
        Worddelay[idles+stages-1,wordlength-2,0] (c1t1) s7 -> s19
        Worddelay[idles+stages-1,wordlength,0] (c1t1) s8 -> s10
        Constant half = stages/2,
            rest = (stages - 1) - half
        Worddelay[half,wordlength-2,0] (c1t0) Xout -> s4
        Worddelay[rest,wordlength-2,0] (c1t0) s4 -> s2
        Bitdelay[wordlength] s1 -> s19
        Bitdelay[wordlength] s19 -> s20
        Bitdelay[wordlength] s20 -> s21
        Bitdelay[wordlength-1] s2 -> s3
        Bitdelay[wordlength-4] s9 -> s11
        Bitdelay[wordlength-4] s10 -> s12
        Bitdelay[wordlength/2+1] s21 -> s22

    END

    ! ALLOCATE VALUES TO PARAMETERS : wordlength,idles,stages

    FIR[14,10,47]

ENDOFPROGRAM
```

Figure 14.7   An example of silicon compilation
             using FIRST.
             (a)   Input system description file

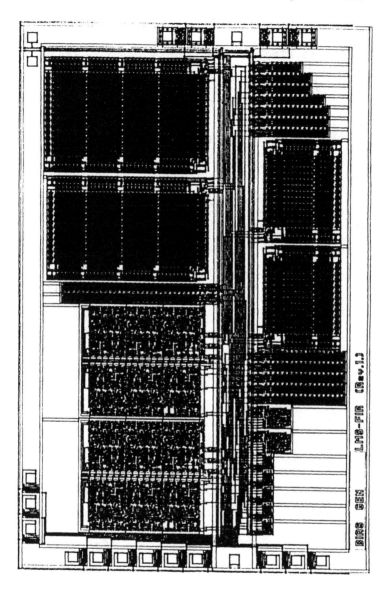

Figure 14.7   (b)   Resulting compiled chip layout
                    (4,000 transistors).

REFERENCES

[1]   Hightower, Design Automation Conference, 1969.

[2]   Werner, J., "Software for Gate-Array Design:
       Who is Really Aiding Whom?", VLSI Design,
       Fourth Quarter, 1981, pp.22-32.

[3]   Locanthi, B., "LAP: A SIMULA Package for IC
       Layout", Caltech Display File 1982, 1978.

[4]   Johannsen, D.L., "Silicon Compilation", Ph.D.
       Thesis, Caltech, 1981.

[5]   Siskind, J.M., Southard, J.R. and Crouch, K.W.,
       "Generating Custom High Performance VLSI Designs
       from Succint Algorithmic Descriptions",
       Proceedings, Conference on Advanced Research
       in VLSI, MIT, 1982.

[6]   Denyer, P.B., Renshaw, D. and Bergmann, N.,
       "A Silicon Compiler for VLSI Signal Processors",
       Proceedings European Solid State Circuits
       Conference (ESSCIRC 82), Brussels, 1982.

Chapter 15

Practical aspects of
semi-custom design

A. Hopper and J. Herbert

15.1   INTRODUCTION

This chapter is intended to augment the formal information on the subject of semi-custom design by giving a flavour of practical experience in this area. In discussing system design (section 15.2), some of the choices and constraints involved are presented and their influence is illustrated with real design examples. It is shown that a solution may be approached directly if the constraints are not severe, but in most practical situations it will be arrived at by a process of iteration where the choices are examined in an interactive way and the designer moves in all directions through the design space.

In the next section we look at CAD tools with particular reference to those which we have used. We see that a CAD system is needed which allows the engineer to make choices and determine constraints such that he can converge on a solution within the limitations of the design and the fabrication process.

15.2   SYSTEM DESIGN

As technology moves forward so it becomes possible for a more and more complex circuit to be implemented in VLSI. At the other end of the spectrum simple circuits become easier to make and thus their price is reduced so that they can be considered for new applications. These factors have led to increased activity in hardware system design.

When a designer approaches a problem some choices and limitations can be specified easily. These may consist of the maximum number of pins available, the maximum gate count or maximum speed of a gate and thus the designer can attempt an implementation rapidly. However, unless very simple circuits are being considered, the direct approach runs out of steam quickly even when the design tools include methods for ensuring that the mapping from the designers drawings to the silicon is direct.

The choices and constraints associated with circuit design and the fabrication process and their influence on hardware system design are discussed below.

15.2.1   Choice of Technology

When considering a design for potential integration we have to make a choice of technology. At present this normally consists of deciding between CMOS and bipolar technologies. Although CMOS takes power when switching and bipolar takes a constant power at all times it does not follow that for large circuits CMOS will be better. For example a ripple counter will in general consume less power in CMOS. If however, for speed reasons, it has to be converted to a synchronous design a bipolar implementation may be better. We can see that already there is a difficulty i.e. a design cannot proceed until some choices are made, while the choices may not be well founded until the design is complete. We would like to explore different choices under various constraints and subsequently compare implementations.

While estimating the power consumption may be an easy calculation, other constraints may be difficult to predict. Consider the following networks in which a signal (S) is fanned-out to four places and gated with another signal (G) (Fig. 15.1). On the assumption that gate delays for the six gates are similar it is easy to arrange that the track delays for the standard technology are similar and thus delays to outputs F0-F3 are matched (Fig. 15.1 (a)). There are two track nets we have to consider, those stemming from G and those from S. In both cases all capacities can be added and we have to make sure the two values are the same.

However, now examine the wire-or technology (Fig. 15.1 (b)). Gates S and G each have a fan-out of 4 and pairs of outputs are wire-ored to give the functions F0-F3. Thus we have a total of 4 independent nets to consider. When delays are dominated by gates (rather than tracking) this may not be a problem but if the reverse is true it may be almost impossible to arrange that the delays to F0-F3 are matched. If for example, we move gate F0 to a distant part of the chip only the delay for that path will increase and we may be constrained in moving F1-F3 to match because of a similar constraint at the next layer and so on. The problem may well be further complicated by the technology not being symmetrical for positive and negative edges - in which case logic functions would always consist of matched pairs of gates. Furthermore, tracking delays may be different in the vertical and horizontal directions making track matching difficult.

This illustrates that with some technologies a simple approach, such as a top down gate array implementation, may be almost impossible to deal with. We can design circuits which do

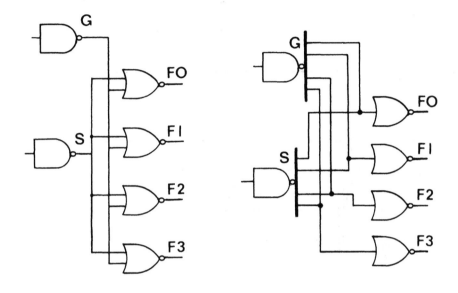

(a) Standard technology   (b) Wire-or technology

Fig. 15.1 Comparison of implementations

not require precise matching but in general this leads to a very conservative approach with a poor ratio of gate count to functionality. To help cope with this problem we would like to have a CAD system which allows us to place constraints on the circuit so that the layout can then be completed in such a way that these are satisfied.

15.2.2   Dealing with Some Constraints

Assuming a technology has been selected, we proceed to try and implement the circuit. Two kinds of difficulties we may encounter are delay constraints for paths and deterioration of signals due to high speed effects. We will illustrate these by considering an actual chip - the graphics controller for the BBC microcomputer. Internally this chip contains a 16 location memory which is addressed using 8 (address) lines (Fig. 15.2).

This is not a particularly difficult circuit to implement for slow speed operation, however, it becomes more difficult if

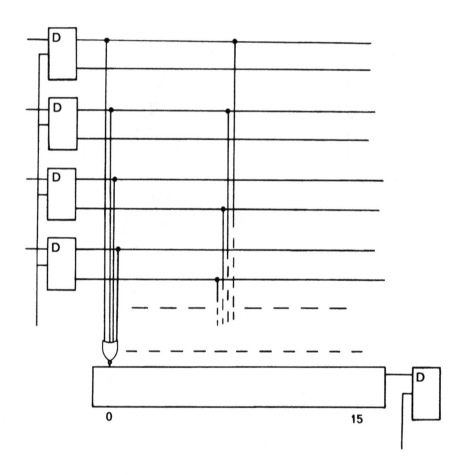

Fig. 15.2 Graphics controller circuit (original design)

either speed or chip size are important. On some gate arrays it may be difficult to route an 8-bit bus without using a lot of silicon area. Furthermore the fan-out on each line may be restricted and thus buffer gates are used. On the BBC machine graphics controller the circuit was altered to that shown in Fig. 15.3 .

Thus the design was changed to meet layout and performance constraints and this could only be done once a partial layout had been attempted. This in turn introduced new constraints; that delays down the address lines were matched so that the correct

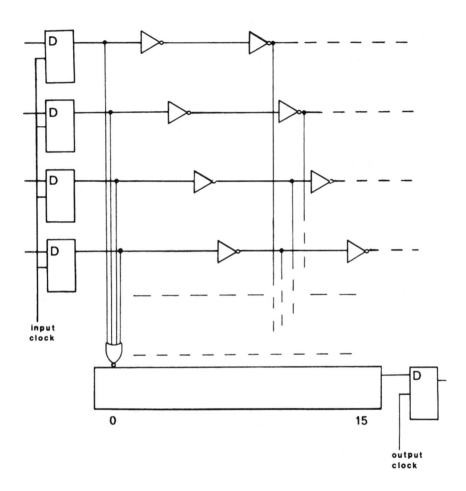

Fig. 15.3 Graphics controller circuit (altered design)

location was decoded both at low and high speeds.

In order for the circuit to operate at full speed (16 MHz) the output clock is a delayed version of the input clock. This delay is implemented using a number of inverters. With a simple 0/1 simulator this delay can be verified and the circuit simulates correctly. In practice however, the delay is a function of both the capacitance of the gate and the time since the last edge. This may mean that consecutive edges run into each other and pulses shrink or even disappear. This effect occurs for bursty or non-symmetrical pulse sequences. Such an effect can result in a circuit not functioning properly and so again a naive approach may fail to find a solution.

This discussion illustrates that care has to be taken to ensure that either the design constraints are very light and therefore a straightforward approach is likely to be successful or that the technology and constraints are well understood and an interactive design automation system can be used to converge on a solution.

15.2.3   Examples of Integrations

Let us now consider two examples where integration has taken place and how this has been achieved.

(a) BBC computer graphics controller.

We have already discussed the graphics processor for the BBC machine. From a user's point of view good graphics is one of the most desirable features in a computer system. It is attractive to provide this in an integrated fashion in the machine. While it would be possible to provide the equivalent functionality using discrete components this would be much more expensive and also would take up a large amount of board space. In particular it is often the case that standard components which have more functionality than is required are used whereas a custom design provides precisely what is needed.

The functions of the BBC graphics controller are to access memory, display the picture and to map logical colours to physical colours by using a palette. The graphics chip also controls the cursor which changes size and position according to the graphics mode being used. As well as these basic features the opportunity is also taken to absorb some further clocking components into the chip which reduces the board size.

The basic technology choice was constrained by the normal considerations of cost, speed and pin count. Because the BBC machine is a consumer product it is important to adopt a low cost technology. First order designs suggested that about 500 gates would be required for implementation, a large proportion operating at the full speed of 16MHz. Thus a relatively fast technology would have to be chosen which at that time suggested Ferranti Current Mode Logic. As pin count could be constrained to 28 pins, packaging was not considered a problem. As we have already discussed, some way into the layout stage the logic design was changed to enable the palette to be completed successfully. The BBC machine is now in use in schools, universities and industry and it arguably has the best graphics of any machine in its price range.

(b) Electron computer ULA.

As another example let us consider the design of a small personal computer system. Such systems normally consist of a microprocessor, some ROM, some RAM, perhaps one or two standard chips and a special chip, the ULA (gate array), which glues the others together and provides special features. This structure is shown below and is the one used for the Acorn Electron Computer (Fig. 15.4).

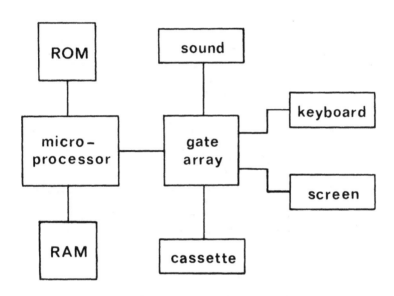

Fig. 15.4 Structure of a small personal computer

By using a special component we are minimising the chip count in the system and thus the cost. At the same time we can take advantage of the flexibility of the semi-custom approach to provide such additional features (sound, serial I/O) as are required. Considering our standard criteria of cost, speed and pin count the most severe constraint is due to pin count. Because the semi-custom chip controls most of the I/O the pin count is high and in the Electron design is 68. Because such large packages are expensive this may mean that the cost of the die may be lower than that of the package. Having established the system architecture we then simulate it in as much detail as possible. In this instance higher level models of the microprocessor (6502), RAM's, ROM's etc.

were built while the gate array was simulated in detail at gate level (2000 gates). This meant that the array was simulated in its correct environment with external parameters being checked against the manufacturers data sheets while internally each gate was simulated down to the detail of exponential rise and fall times. The actual time to do this was about two months although the elapsed time was longer because the design was changed as a result of simulation results. The system is working successfully and because the database exists in the design automation system we can easily consider absorbing more components onto a future semi-custom or perhaps custom chip.

15.3   CAD TOOLS

The single most important feature of a gate array CAD package is that it provides a completely integrated environment for the designer. A package of this type has been designed at the University of Cambridge Computer Laboratory (1). The various parts operate on one database, with clean interfaces between stages, so that the designer can easily progress his circuit from an early conceptual stage through to a working device. The package was designed with Ferranti CML technology in mind and this means that some algorithms are particularly targeted for features of that system (for example single layer metalisation). Wherever possible, general solutions were adopted and the system has been structured with clean interfaces to implementation specific details. The system can therefore be expanded to incorporate arrays from other manufacturers. Some aspects of this CAD system are discussed below.

15.3.1   Circuit Description

While most hardware description languages at present are targeted for the engineer our approach is to provide a system equally suitable for the programmer (2). The design of sophisticated chips is becoming increasingly a question of software development. The design is described using a program, is tested by running it in conjunction with other programs, the program then adjusted until the design is considered correct. Powerful techniques have been developed by computer scientists to enable complex programs to be written. Such techniques are now applicable to chip design and this saves us having to design special hardware description languages and build the usual tools (e.g. compiler, debugger). A skilled designer may use the full power of the programming language to describe his circuit, structuring the program to reflect the physical and/or functional hierarchy accurately, using repeat features to describe regular structures and so on. Where the programming talent of the designer is limited he may write his description as a long sequence of simple statements and perhaps use some low level macros provided for him in a library. In either case the circuit description is then compiled, with syntactic and semantic checking performed by the computer, and the compiled program forms a part of the design database.

15.3.2   Simulation and Layout

When simulating the design we wish to provide input waveforms, monitor points within the circuit and external pins and by doing so verify the design. The simulator should be fast and capable of accepting complex input patterns, perhaps themselves emanating from higher level models of the environment rather than simply as a series of changes on pins. This is accomplished in the CAD system by describing the driving waveforms using a program. These can thus be complex, conditional, random, or represent any other whim of the designers imagination. Specifically, sequences of complex input patterns can be called from a higher level program making driving of the simulator easier.

When designing complex systems the gate array may form only a small part of the complete system. In particular it may communicate with other chips on the board. The simulator allows other components of this type to be incorporated, using information much the same as in the manufacturers data sheet, so that the gate array is modelled in the correct environment. In general the simulator is very interactive, can display waveforms on a graphics terminal, simulate forwards or backwards, change scales and signals rapidly. While the simulation is taking place gates are modelled as changing after a delay and initially are set to a value X (unknown). However in many cases, particularly at high speeds, this is insufficient because gates switch exponentially and we may observe pulse shrinkage as a result. The simulator models the exponential rise and fall times and adjusts the delays of gates with no significant loss in speed.

Conventionally the layout is done after the simulation has been completed. However with more complex, and in particular higher speed systems the layout forms an integral part of the design process and will be done in parallel with the simulation.

As we simulate, the layout may also be in progress and some initial partitioning might have taken place. Thus it may be evident at an early stage that some tracks are likely to be longer than others and the simulator adjusts estimates of gate delays as this information becomes available. Initially we require a system which will partition the design (either manually or automatically) into modules and enable us to move these on the surface of the chip. Congestion is computed and alleviated. Once the layout has been done at one level we may move up a level or down a level and repeat the process. In our system the layout software is targeted for Ferranti ULAs and while the system for dealing with modules on the surface of the chip is general once we move down to the chip level gates and interconnections are array specific. In particular Ferranti ULAs do not provide channels for routing tracks between modules, instead long tracks have to be routed through other logic.

It can be seen that the combination of an integrated system, a powerful simulator, hierarchical layout and interaction allows a designer to converge on a solution which is both efficient in the use of silicon and verifies the design in detail. Once this has been done a tape can be produced suitable for manufacture or for direct commitment using electron beam techniques. Thus the time to design and prototype a chip can be very low.

15.3.3   CAD System Implementation

The CAD system described above has been implemented on a M68000 microcomputer with 1 or 2 megabytes of main memory. A machine of this type is sufficiently powerful to simulate and compute layouts in reasonable times and interactions take place within the 'boredom time' of the user. Experiments show that because the algorithms used in chip design tend to be CPU bound we can make a machine of this class run at a speed comparable to a single user VAX-750 (though at a much lower cost). Thus the provision of a large number of workstations for design engineers is a practical proposition.

One of the most positive results obtained when using the M68000 systems was the marked acceleration of progress when several workstations were in use at one time. The M68000 systems are connected to a Cambridge Ring network. The file system is provided by a single network file server and thus the workstations share the same file space. When simulating the designer often wishes to run the same simulation many times with only minor parameter changes. By using several workstations he can initiate a number of simulations at the same time. While this is taking place he may be editing other files or inspecting results. In the current Cambridge Ring environment there are 15 workstations, however a typical engineer can only drive about 6 before losing track of events. Even with only 6 in use the volume of work being done is equivalent to a large mainframe computer.

15.3.4   Extensions to the Current System

The next stage is to adapt the system to cell based gate arrays. This is an easy job at the simulation stage and only involves incorporation of the appropriate macros in a library suite. The simulator design allows this kind of extension. A completed chip design would then be passed over to the manufacturer in a digital form, specifying cells from the library of that manufacturer as required. To make the system useful at an early stage a reverse path would be provided to enable details of delays and layout to be incorporated in the simulation. While this means the layout is done on another system a design can be completely tested on the workstation.

To make the CAD system capable of handling a wider range of arrays and in due course 'super chips' (designs with a micro, memory and gate array on one chip), a more fundamental redesign is required. While this is not likely to comprise large changes, careful thought would ensure that a proper front-end, back-end approach could be adopted. This would allow a design to be explored, perhaps a floor plan completed, independent of the fabrication process. Once the technology was known the chip could be implemented using the verified database. To allow this several extensions would be provided (transmission gates, tri-state in the simulator; channel routing, custom at the layout stage) and these would work within the single integrated software environment. It would be particularly attractive if an engineer while developing a design (for example comprising a micro, some memory and some logic) could simulate it in a front-end; while trying a number of options in the back-end ranging from several gate arrays to a single superchip. The trade-off of different options can then be accurately assessed.

15.3.5   Associated Research

At the University of Cambridge Computer Laboratory research based on the work of Gordon (3) is being conducted into the formal verification of a circuit design. This involves proving formally that an implementation of a circuit agrees with the specification of that circuit. A system has been built which deduces the formal behaviour of a circuit from the implementation as given by the circuit description. (This is the same circuit description which provides the database for simulation and layout). This behaviour and a specification of the circuit are loaded into a system (Cambridge LCF) in which proofs may be conducted. Cambridge LCF is a proof assistant and can be used interactively or in batch fashion. It gives the user great freedom in the strategies he may employ when conducting a proof. Essentially, the formula equating the specification to the implementation is transformed until it is proved to be true or false. Examples have already been verified in this manner.

The model used at present in formal verification caters for synchronous circuits and ignores the influence of timing. An addition to the verification system will prove that timing constraints are fulfilled and thus justify this simplification. Furthermore, it will be possible to specify the timing of interface signals with respect to the basic clock. This will allow the designer to partition the design in the time domain. A new formal model is envisaged which will incorporate timing. This will then provide a unique basis for the more complete specification and verification of circuits.

Formal verification will be of great practical use when it can be provided as a relatively painless tool for the hardware designer. It will eliminate much of the drudgery in simulation and allow more time to be spent on the subtle problems.

15.4   SUMMARY

Some practical experience in the field of semi-custom design has been discussed in this chapter. The design process has been shown to be influenced at present by the choices available and the constraints. A direct approach will succeed only if the constraints are light. In general we wish to explore the design space, determining constraints and examining choices so that we can converge on a solution. We have seen the advantages of having a flexible, integrated CAD system which allows partitioning of the design and feedback from the layout to simulation.

Lighter constraints have not necessarily followed from technological advances. For example, a decrease in feature size may increase the influence of tracking delay. Semi-custom design has inherently more constraints than full custom design. We have described a CAD system in use at present where the designer can determine constraints and explore various design options manually. An ideal system would be able to take a specification, evaluate different choices and produce an optimum implementation automatically. Such an ideal solution may come from silicon compilers (discussed in chapter 14) which synthesise a chip from a high level description. These may in the long term take over much of the present work of the hardware designer.

In the immediate future we envisage the increased use of design disciplines and implementation technologies which free the designer from lower level constraints and allow for a safer implementation. Employing a design discipline at a high level should aid a structured approach to the problem and allow many lower level constraints to be ignored. A restriction to synchronous circuits for instance eases timing problems and simplifies the formal verification of the design (cf. 15.3.5.). The use of self-timed logic as a means of precluding timing problems has been proposed (4). The ASM method of logic design (cf. chapters 7 and 9) allows a design to be developed at a level of abstraction above the circuit description. By submitting to design disciplines such as these a designer can follow a much safer design path.

It is evident from the examples presented earlier in this chapter that the method of implementation can influence greatly the ease of achieving a correct solution. Various methods of implementation including gate arrays, PLA, PAL and cell-based systems have been treated in other chapters. The method of implementation may neccessitate some design discipline to ensure that the mapping from the circuit to silicon is successful. It is desirable that the means of implementation should have as little influence as possible on the higher level design.

In conclusion it may be hoped that many of the practical problems mentioned here will in the future be eliminated or avoided. The problems associated with the design task may then be confined to a higher level and the hardware designer may be free to tackle these and indulge in more creative work.

REFERENCES

1. Robinson, P. and Dion, J., 1982, 'Design Aids for Uncommitted Logic Arrays', Proceedings of the 2nd International Conference on semi-custom ICs.

2. Robinson, P. and Dion, J., 1983 'Programming Languages for Hardware Description', Proceedings of the 20th Design Automation Conference pp.12-16.

3. Gordon, M., 1981, 'A model of register transfer systems with applications to microcode and VLSI correctness', Internal report CSR-82-81, Department of Computer Science, University of Edinburgh.

4. Seitz, C. L., 1980 'System timing', Introduction to VLSI systems (Mead, C. and Conway, L.) chapter 7.

Index

Printed in the USA
CPSIA information can be obtained
at www.ICGtesting.com
JSHW011509221024
72173JS00005B/1260

9 780863 410116